AF576019

Urban Atmospheric Aerosols

Urban Atmospheric Aerosols

Sources, Analysis and Effects

Editors

Regina M. B. O. Duarte
Armando da Costa Duarte

MDPI • Basel • Beijing • Wuhan • Barcelona • Belgrade • Manchester • Tokyo • Cluj • Tianjin

Editors

Regina M. B. O. Duarte
Department of Chemistry and
CESAM – Centre for
Environmental and Marine
Studies, University of Aveiro
Portugal

Armando da Costa Duarte
Department of Chemistry and
CESAM – Centre for
Environmental and Marine
Studies, University of Aveiro
Portugal

Editorial Office

MDPI
St. Alban-Anlage 66
4052 Basel, Switzerland

This is a reprint of articles from the Special Issue published online in the open access journal *Atmosphere* (ISSN 2073-4433) (available at: https://www.mdpi.com/journal/atmosphere/special_issues/Urban_Aerosols).

For citation purposes, cite each article independently as indicated on the article page online and as indicated below:

LastName, A.A.; LastName, B.B.; LastName, C.C. Article Title. *Journal Name* **Year**, *Volume Number*, Page Range.

ISBN 978-3-03943-931-7 (Hbk)
ISBN 978-3-03943-932-4 (PDF)

Contents

About the Editors

Regina M. B. O. Duarte is Principal Researcher at the Department of Chemistry and CESAM - Centre for Environmental and Marine Studies, at University of Aveiro (Portugal). She obtained a degree in Analytical Chemistry (1998), an MSc in Sciences of the Coastal Zone (2001), and a PhD in Chemistry (2006), all at the University of Aveiro. In 2018, Regina Duarte obtained a Fulbright Grant for Portuguese Scholars and Researchers (Advanced Research and University Lecturing Award in the United States) to develop research at Old Dominion University (Norfolk, VA, USA) in the application of advanced multidimensional analytical techniques for the molecular characterization of the complex organic air particles. Her research is focused on the development of new multidimensional analytical research strategies that are aimed at understanding the structural features of organic air particles. Her research interests also include the health effects of organic aerosols, as well as wet and dry deposition of organic aerosols. She has coordinated/participated in several multidisciplinary research projects related to the characterization of complex organic matrices from atmospheric particles. She has authored/co-authored more than 45 peer-reviewed publications.

Armando C. Duarte has been a Professor of Environmental & Analytical Chemistry at the University of Aveiro (Portugal) since 1995, leader of a research group on the same subject, lecturer on Analytical Quality Control subjects, and either supervising or co-supervising many PhD students, some of whom became members of staff, both as lecturers and researchers. Armando Duarte graduated in Chemical Engineering (1977) at the University of Oporto (Portugal) and obtained a PhD in Public Health Engineering (1981) at the University of Newcastle-upon-Tyne (United Kingdom). In 2006, The Portuguese Science and Technology Foundation (FCT) awarded him a prize for Scientific Excellence, and in 2013, he became a Member of the FCT Scientific Council for Natural and Environmental Sciences. His highly-interdisciplinary research spans areas of comprehensive environmental and analytical chemistry, qualimetrics, and analytical quality assurance, but also includes the assessment of the relevance of new concepts and integration of different ideas into widely accepted frameworks, especially when applied to sustainability. He has co-authored more than 470 peer-reviewed publications and edited several books.

Preface to "Urban Atmospheric Aerosols"

Atmospheric fine particulate matter (PM2.5, diameter less than 2.5 μm) has profound effects on radiative climate forcing, atmospheric chemistry, air quality, visibility, and human health. PM2.5 can be directly emitted into the atmosphere (primary aerosols) or formed in the atmosphere (secondary aerosols) through gas-to-particle conversion processes of gaseous species. Furthermore, primary and secondary aerosols may also undergo chemical and physical transformations as they are subjected to transport, atmospheric aging, and removal from the atmosphere. The physical and chemical characterization of PM2.5, its source apportionment, and the assessment of the magnitude and distribution of its emissions are crucial for establishing effective fine air particle regulations and assessing the associated risks to human health.

Urban locations are a very special case as far as PM2.5 concentrations, compositions, sources, and health effects are concerned. Understanding the physical and chemical properties of urban PM2.5 (e.g., atmospheric concentration, size distribution, surface area, chemical composition, and water solubility) is essential for better predicting the tight connection between PM2.5 and its contribution to atmospheric chemistry and the health of both human populations and environmentally sensitive ecosystems. Therefore, it is highly beneficial to conduct studies on the physico-chemical characteristics and toxicological effects of urban PM2.5 in order to establish efficient control strategies. Furthermore, understanding how urban aerosols affect the air quality of indoor environments in urban buildings is also essential for assessing the potential health effects.

This book emerged from the Special Issue "Urban Atmospheric Aerosols: Sources, Analysis, and Effects", published in *Atmosphere*. The papers presented in this book highlight some important aspects concerning the chemical characteristics, optical properties, size distribution, sources, and potential health effects of urban air particles. These studies will be of interest to the atmospheric research community, including those interested in outdoor and indoor air quality, air particle toxicity, and atmospheric chemistry, as well as global climate modelers.

The Editors would like to thank the authors who generously contributed their time and knowledge to ensure the high quality of this work. The Editors also express their gratitude to the *Atmosphere* journal editors, reviewers, and the production team for their invaluable support and cooperation in the publication of the book. The Editors are also deeply grateful to FCT/MCTES for the financial support to CESAM (UIDP/50017/2020+UIDB/50017/2020), AMBIEnCE project (PTDC/CTA-AMB/28582/2017), and the Exploratory Research Project (IF/00798/2015/CP1302/CT0015). The Surface Ocean-Lower Atmosphere Study (SOLAS) is also acknowledged for endorsing the AMBIEnCE project. We sincerely hope that this work will stimulate further development to improve the current understanding of the sources, composition, fate, and impact of urban air particles.

Regina M. B. O. Duarte, Armando da Costa Duarte
Editors

Editorial

Urban Atmospheric Aerosols: Sources, Analysis, and Effects

Regina M. B. O. Duarte * and Armando C. Duarte

Department of Chemistry and CESAM—Centre for Environmental and Marine Studies, University of Aveiro, 3810-193 Aveiro, Portugal; aduarte@ua.pt

* Correspondence: regina.duarte@ua.pt

Received: 9 November 2020; Accepted: 10 November 2020; Published: 12 November 2020

Keywords: urban atmospheric aerosols; primary organic aerosols; secondary organic aerosols; chemical composition; toxic elements; particle size characterization; urban air quality; indoor air pollution

---

Atmospheric fine particulate matter ($PM_{2.5}$, aerodynamic diameter less than 2.5 μm) has profound effects on radiative climate forcing, atmospheric chemistry, air quality and visibility, and human health [1,2]. $PM_{2.5}$ can be directly emitted into the atmosphere (primary aerosols) from a diversity of natural and anthropogenic sources, including biomass burning, the incomplete combustion of fossil fuels, volcanic eruptions, and wind-driven or traffic-related suspension of road, soil, and mineral dust, sea salt, and biological materials [1]. Nonetheless, $PM_{2.5}$ can be also formed in the atmosphere (secondary aerosols) through gas-to-particle conversion processes of gaseous species (i.e., nucleation, condensation, and heterogeneous and multiphase chemical reactions) [1,3]. Furthermore, primary and secondary aerosols may undergo chemical and physical transformations, being subjected to transport, cloud processing, and removal from the atmosphere [4]. This multitude of emission sources and formation/processing mechanisms contribute to the diversity and complexity of the chemical composition and physical properties (i.e., concentration, size distribution, and surface area) of atmospheric aerosols, which in turn influences the climate and health effects of atmospheric fine air particle, further adding a layer of complexity. Indeed, the physical and chemical characterization of $PM_{2.5}$, its source apportionment, and the assessment of the magnitude and distribution of $PM_{2.5}$ emissions is crucial for establishing effective fine particulate matter regulations and assessing the associated risks to human health.

Due to ever increasing urbanization, urban areas are a very special case as far as $PM_{2.5}$ concentrations, composition, sources, and health effects are concerned. A recent study conducted in five major cities (Athens and Paris in Europe, Pittsburgh and Los Angeles in the United States, and Mexico City in Central America) showed that most of the fine particulate matter (PM_1 or $PM_{2.5}$) is secondary (between 50% and 75% of aerosol mass load) [5]. The secondary components include sulfates and ammonium nitrate, but also oxidized organic compounds, whereas the major primary components are elemental (or black) carbon, metal oxides and crustal material, and fresh organic particulate matter. According to Pandis et al. [5], the importance of local sources is also quite different depending on the studied aerosol chemical component. For primary components such as elemental carbon, the local sources dominate; however, the situation is very different for secondary components, with most of the secondary material being due to either medium- or long-range sources [5]. Under this intricate scenario, knowledge of the detailed chemical composition, physical properties, and sources of fine particulate matter is crucial for assessing and managing the air quality surrounding large urban centers of both developed and developing countries. Nowadays, it is also well established that biological responses to $PM_{2.5}$ go beyond the particulate matter mass [6], and that additional air quality metrics (e.g., black carbon, secondary organic and inorganic aerosols) may be valuable in evaluating the health risks of $PM_{2.5}$ at urban locations (e.g., [7]). Furthermore, understanding how outdoor $PM_{2.5}$ in urban areas affects the air quality of indoor environments in urban buildings and houses is also essential for assessing the potential health effects on residents [8,9]. Hence, much work is still needed to enhance our understanding of the chemical composition, size distribution, source apportionment, and indoor–outdoor relationships of $PM_{2.5}$ in urban areas and their health consequences upon exposure. Additional studies are also necessary to establish a more comprehensive understanding of how emission reduction measures will affect the levels and characteristics of $PM_{2.5}$ at urban locations.

The Special Issue "Urban Atmospheric Aerosols: Sources, Analysis, and Effects" is motivated by the need to address some important aspects concerning the chemical characteristics, optical properties, size-distribution, sources, and potential health effects of urban air particles. It comprises seven peer-reviewed, open access articles spanning the main topics of the field. For example, Hussein et al. [10] explore the temporal variability (diurnal, weekly, and seasonal) of particle number size distribution in the city of Amman (Jordan), as an example of an urban Middle Eastern environment. This study shows different seasonal variations for the submicron and coarse mode particle number concentrations. The authors also suggest the contribution of new particle formation for the submicron aerosols fraction, whereas sand and dust storms are the most important sources of coarse mode aerosols. These data highlight that the urban Middle Eastern region under investigation is exposed to the impacts of aerosols from various sources, which could become a serious health hazard if persistent for long periods of time. In another study, Li et al. [11] address the impacts of emission reduction measures on the characteristics and sources of trace elements in $PM_{2.5}$ at urban and suburban areas of Beijing, both before and after the 2014 Asia-Pacific Economic Cooperation (APEC) summit, which was held in Beijing. According to the authors, the air quality improved during APEC, with toxic elements, such as vanadium (V), chromium (Cr), manganese (Mn), arsenic (As), cadmium (Cd), and lead (Pb), decreasing more than 40% due to the emission regulations. Li et al. [11] highlight that future control efforts for toxic elements in megacities, such as Beijing, should focus on coal and oil combustion as well as on traffic emissions. Nevertheless, assessing the potential environmental exposure effects to atmospheric pollution (outdoor and indoor) is also important at locations other than megacities. In the study of Moreda-Piñeiro et al. [12], the authors assess the levels of particle-bound mercury (PHg) in PM_{10} (particulate matter with aerodynamic diameter less than 10 μm) at several sites (industrial, urban, and suburban) of an Atlantic coastal European region (i.e., the northwest of Spain) over 13 months. This study shows that the levels of PHg varied between 1.5–30.8, 1.5–75.3, and 2.27–33.7 pg m^{-3} at urban, suburban, and industrial sites, respectively. Data analysis also suggested an anthropogenic origin of PHg at the urban site, whereas biomass burning was likely to be the main source of PHg at the suburban site [12]. A sea salt and crustal/anthropogenic origin of PHg is suggested at the industrial site [12]. A toxicity estimate of PHg, using hazard quotient, suggested no non-carcinogenic risk for adults at the studied sites [12]. On the other hand, the preliminary study of Vicente et al. [13] provides a first insight into the occurrence of plasticizers and polycyclic aromatic hydrocarbons (PAHs) in PM_{10} from resuspended dust samples in Spanish households. The authors suggest that exposure to plasticizers and PAHs through the ingestion route poses much higher risks as compared to inhalation and dermal contact, which is of particular concern for infants due to their higher dust intake via hand-to-mouth activities. For the sake of easier comparison between different locations/houses, the authors also recommend the development of standard operational procedures for household dust sampling and analysis [13].

In the context of atmospheric pollution and global climate change, another important aspect is the need to unravel the chemical characteristics of atmospheric air particulate matter. Among the various $PM_{2.5}$ components, the water-soluble organic nitrogen (WSON) is the least studied aerosol component, although its role in secondary organic aerosol formation and its toxicity (e.g., [14] and references therein) has long been recognized. In this regard, the concentration of the WSON, with a particular focus on its seasonal variations and potential sources, is highlighted in the study of Chen et al. [15], which was carried out in urban Yangzhou (eastern China). This study shows that the annual mean WSON concentration (1.71 ± 1.08 μg m^{-3}) in Yangzhou was much higher than those in Japan, Greece, and Florida (USA). It is also shown that the WSON concentrations had no significant seasonal and monthly variation; however, the WSON-to-total nitrogen mass ratios were much higher in summer and autumn than in winter and spring. The authors concluded that secondary formation is an important source of aerosol WSON in Yangzhou.

An additional approach to monitoring air pollution in urban and heavily industrialized areas is the consideration of aerosol optical depth (AOD), which measures the light extinction by aerosol scattering and absorption in the atmospheric column. Xue et al. [16] applied AOD data acquired from Moderate Resolution Imaging Spectroradiometer (MODIS) to assess whether correlations could be established between AOD and urban development, construction factors, and geographical environment factors in the Shandong Province over a 10-year timeframe (2007 to 2017). It was concluded that the introduction of targeted environmental protection policies was effective in alleviating pollution-related problems in the process of urbanization. Nonetheless, proper sampling of satellite AOD data is extremely important for a correct characterization of aerosol properties and air pollution over the areas of interest. Tian and Gao [17] address this challenge by assessing the accuracy of Terra-MODIS AOD retrieval products using the ground observations from twenty-three AErosol RObotic NETwork (AERONET) sites in heavy aerosol loading areas of Asia (including China, India, and Pakistan), the Middle East (including the United Arab Emirates, Bahrain, and Kuwait), and Northern Africa (including Egypt, Niger, Benin, Mali, and Nigeria). This study highlights the impact of land cover type and seasonal variation in the accuracy of the MODIS aerosol products and emphasizes the need for a suitable choice of the MODIS aerosol algorithms and products to assure a proper prediction of the air particles—AOD relationship.

In summary, the studies and the results discussed in this Special Issue will be of interest to the atmospheric research community, including those interested in air quality outdoors and indoors, visibility, the toxicity, composition, and sources of air particles, atmospheric chemistry, as well as global climate modelers. Finally, it is expected that the research presented here will inspire new research questions and hypotheses, which will help to untangle the strong connections between air particles and their impact on atmospheric chemistry, climate, and human health.

Author Contributions: Conceptualization, investigation, writing—original draft preparation, review and editing, funding acquisition: R.M.B.O.D. and A.C.D. Both authors have read and agree to the published version of the manuscript.

Funding: Thanks are due to FCT/MCTES for the financial support to CESAM (UIDP/50017/2020+UIDB/50017/2020) through national funds. This work was supported by project AMBIEnCE (PTDC/CTA-AMB/28582/2017) funded by FEDER, through COMPETE2020-Programa Operacional Competitividade e Internacionalização (POCI), and by national funds (OE), through FCT/MCTES. Regina M.B.O. Duarte also acknowledges financial support from FCT/MCTES through an Exploratory Research Project (IF/00798/2015/CP1302/CT0015, Investigator FCT Contract IF/00798/2015). Surface Ocean-Lower Atmosphere Study (SOLAS) is also acknowledged for endorsing the AMBIEnCE project.

Acknowledgments: We, the guest editors of this Special Issue of *Atmosphere*, are grateful to all of the authors, reviewers, and MDPI staff. We hope that the papers collected here will stimulate further development to improve the current understanding on the sources, analysis, and effects of urban atmospheric aerosols.

Conflicts of Interest: The authors declare no conflict of interest.

References

1. Pöschl, U. Atmospheric Aerosols: Composition, Transformation, Climate and Health Effects. *Angew. Chem. Int. Ed.* **2005**, *44*, 7520–7540. [CrossRef] [PubMed]
2. Heal, M.R.; Kumar, P.; Harrison, R.M. Particles, air quality, policy and health. *Chem. Soc. Rev.* **2012**, *41*, 6606. [CrossRef] [PubMed]
3. Hallquist, M.; Wenger, J.C.; Baltensperger, U.; Rudich, Y.; Simpson, D.; Claeys, M.; Dommen, J.; Donahue, N.M.; George, C.; Goldstein, A.H.; et al. The formation, properties and impact of secondary organic aerosol: Current and emerging issues. *Atmospheric Chem. Phys. Discuss.* **2009**, *9*, 5155–5236. [CrossRef]
4. Zhang, R.; Wang, G.; Guo, S.; Zamora, M.L.; Ying, Q.; Lin, Y.; Wang, W.; Hu, M.; Wang, Y. Formation of Urban Fine Particulate Matter. *Chem. Rev.* **2015**, *115*, 3803–3855. [CrossRef] [PubMed]

5. Pandis, S.N.; Skyllakou, K.; Florou, K.; Kostenidou, E.; Kaltsonoudis, C.; Hasa, E.; Presto, A.A. Urban particulate matter pollution: A tale of five cities. *Faraday Discuss.* **2016**, *189*, 277–290. [CrossRef] [PubMed]
6. Cassee, F.R.; Héroux, M.-E.; Gerlofs-Nijland, M.E.; Kelly, F.J. Particulate matter beyond mass: Recent health evidence on the role of fractions, chemical constituents and sources of emission. *Inhal. Toxicol.* **2013**, *25*, 802–812. [CrossRef] [PubMed]
7. Almeida, A.S.; Ferreira, R.M.P.; Silva, A.M.S.; Duarte, A.C.; Neves, B.M.; Duarte, R.M.B.O. Structural Features and Pro-Inflammatory Effects of Water-Soluble Organic Matter in Inhalable Fine Urban Air Particles. *Environ. Sci. Technol.* **2020**, *54*, 1082–1091. [CrossRef] [PubMed]
8. Leung, D.Y.C. Outdoor-indoor air pollution in urban environment: Challenges and opportunity. *Front. Environ. Sci.* **2015**, *2*, 1–7. [CrossRef]
9. Zhao, L.; Chen, C.; Wang, P.; Chen, Z.; Cao, S.; Wang, Q.; Xie, G.; Wan, Y.; Wang, Y.; Lu, B. Influence of atmospheric fine particulate matter (PM2.5) pollution on indoor environment during winter in Beijing. *Build. Environ.* **2015**, *87*, 283–291. [CrossRef]
10. Hussein, T.; Dada, L.; Hakala, S.; Petäjä, T.; Kulmala, M. Urban aerosol particle size characterization in Eastern Mediterranean Conditions. *Atmosphere* **2019**, *10*, 710. [CrossRef]
11. Li, M.; Liu, Z.; Chen, J.; Huang, X.; Liu, J.; Xie, Y.; Hu, B.; Xu, Z.; Zhang, Y.; Wang, Y. Characteristics and source apportionment of metallic elements in PM2.5 at urban and suburban sites in Beijing: Implication of emission reduction. *Atmosphere* **2019**, *10*, 105. [CrossRef]
12. Moreda-Piñeiro, J.; Rodríguez-Cabo, A.; Fernández-Amado, M.; Piñeiro-Iglesias, M.; Muniategui-Lorenzo, S.; López-Mahía, P. Levels and sources of atmospheric particle-bound mercury in atmospheric particulate matter (PM10) at several sites of an Atlantic coastal European region. *Atmosphere* **2020**, *11*, 33. [CrossRef]
13. Vicente, E.D.; Vicente, A.; Nunes, T.; Calvo, A.; del Blanco-Alegre, C.; Oduber, F.; Castro, A.; Fraile, R.; Amato, F.; Alves, C. Household Dust: Loadings and PM10-Bound plasticizers and polycyclic aromatic hydrocarbons. *Atmosphere* **2019**, *10*, 785. [CrossRef]
14. Cornell, S.E. Atmospheric nitrogen deposition: Revisiting the question of the importance of the organic component. *Environ. Pollut.* **2011**, *159*, 2214–2222. [CrossRef]
15. Chen, Y.; Chen, Y.; Xie, X.; Ye, Z.; Li, Q.; Ge, X.; Chen, M. Chemical characteristics of PM2.5 and water-soluble organic nitrogen in Yangzhou, China. *Atmosphere* **2019**, *10*, 178. [CrossRef]
16. Xue, R.; Ai, B.; Lin, Y.; Pang, B.; Shang, H. Spatial and temporal distribution of aerosol optical depth and its relationship with urbanization in Shandong province. *Atmosphere* **2019**, *10*, 110. [CrossRef]
17. Tian, X.; Gao, Z. Validation and accuracy assessment of MODIS C6.1 aerosol products over the heavy aerosol loading area. *Atmosphere* **2019**, *10*, 548. [CrossRef]

Publisher's Note: MDPI stays neutral with regard to jurisdictional claims in published maps and institutional affiliations.

Article

Urban Aerosol Particle Size Characterization in Eastern Mediterranean Conditions

Tareq Hussein [1,2,*], Lubna Dada [2], Simo Hakala [2], Tuukka Petäjä [2] and Markku Kulmala [2]

1 Department of Physics, The University of Jordan, Amman 11942, Jordan
2 Institute for Atmospheric and Earth System Research (INAR), University of Helsinki, PL 64, FI-00014 UHEL Helsinki, Finland; lubna.dada@helsinki.fi (L.D.); simo.hakala@helsinki.fi (S.H.); tuukka.petaja@helsinki.fi (T.P.); markku.kulmala@helsinki.fi (M.K.)
* Correspondence: tareq.hussein@helsinki.fi

Received: 13 October 2019; Accepted: 10 November 2019; Published: 14 November 2019

Abstract: Characterization of urban particle number size distribution (*PNSD*) has been rarely reported/performed in the Middle East. Therefore, we aimed at characterizing the *PNSD* (0.01–10 μm) in Amman as an example for an urban Middle Eastern environment. The daily mean submicron particle number concentration (PN_{Sub}) was 6.5×10^3–7.7×10^4 cm^{-3} and the monthly mean coarse mode particle number concentration (PN_{Coarse}) was 0.9–3.8 cm^{-3} and both had distinguished seasonal variation. The PN_{Sub} also had a clear diurnal and weekly cycle with higher concentrations on workdays (Sunday–Thursday; over 3.3×10^4 cm^{-3}) than on weekends (below 2.7×10^4 cm^{-3}). The PN_{Sub} constitute of 93% ultrafine fraction (diameter < 100 nm). The mean particle number size distributions was characterized with four well-separated submicron modes ($D_{pg,I}$, N_i): nucleation (22 nm, 9.4×10^3 cm^{-3}), Aitken (62 nm, 3.9×10^3 cm^{-3}), accumulation (225 nm, 158 cm^{-3}), and coarse (2.23 μm, 1.2 cm^{-3}) in addition to a mode with small geometric mean diameter (GMD) that represented the early stage of new particle formation (NPF) events. The wind speed and temperature had major impacts on the concentrations. The PN_{Coarse} had a U-shape with respect to wind speed and PN_{Sub} decreased with wind speed. The effect of temperature and relative humidity was complex and require further investigations.

Keywords: submicron; coarse; modal structure; meteorological effect; particle number size distribution; seasonal; diurnal

1. Introduction

Atmospheric aerosol particles have gained increased attention during recent years due to their effects on the climate and human health [1,2]. While most health studies related to air pollution focus on the mass accumulation of particulate matter (PM), advanced studies have shown the importance of the number size distribution on health. For instance, it has been evident that fine particles are capable of penetrating through the respiratory system and circulating in the bloodstream causing serious health effects in different organs in the human body [3]. While the smallest particles do not contribute significantly to the total measured mass concentration, they constitute the highest fraction of the number concentration of measured particles in various locations, which can be more than 100,000 particles/cm^3. Besides that, governmental policies tend to enforce strategies that eradicate the total mass in an attempt of improving visibility and saving human health; at the same time ignoring the effects of the ultrafine particles (UFP, diameter < 0.1 μm), which have rather complex sources and atmospheric processes. For example, fine and UFP are capable of growing to reach sizes where they constitute a fraction of cloud condensation nuclei (CCN); and thus, indirectly affecting the climate [4,5].

The drivers behind aerosol particles vary between natural and anthropogenic as well as primary and secondary. Primary particles are emitted to the atmosphere as particles while secondary particles form in the atmosphere via gas-to-particle transformation, which has been known as new particle formation (NPF) observed in various environments and contributing to a major fraction of the total particle number budget [6–8]. Indeed, the complexity of urban aerosols lies in the fact that several sources can contribute in the same particle size range, making it difficult to entangle [9,10]. Although, some studies focusing on the particle number size distribution have been conducted around the East Mediterranean region at both urban and remote sites [11–14], very few studies can be found regarding the Middle East [15–21]. In fact, the Middle East and North Africa (MENA) is of extreme interest as it serves as a compilation of aerosol particle sources including natural dust, anthropogenic pollution from petrochemical industry and urbanization, as well as new particle formation.

The aim of this paper is to characterize the aerosol particle number size distribution and explore the drivers behind diurnal and seasonal variability of the number concentrations in Amman, which is an example of Middle Eastern urban conditions. Here we utilized a long-term measurement of particle number size distributions over a wide particle diameter range (0.01–10 µm). We also applied modal structure analysis for the population of aerosols by using the multi-lognormal distribution function. We additionally investigated the effect of local meteorology on the aerosol concentrations.

2. Materials and Methods

2.1. Aerosol Measurements

The long-term aerosol measurement was performed during 1 August 2016–31 July 2017 at the Aerosol Laboratory, which was located on the third floor of the Department of Physics, University of Jordan. The University of Jordan campus was located at an urban background in the north part of Amman, Jordan (Supplementary Material Figure S1). The aerosol measurement consisted of total number concentration and particle number size distribution, described in more detail in the following section.

The particle number size distribution was measured with a scanning mobility particle sizer (NanoScan SMPS 3910, TSI, Minnesota, U.S.) and an optical particle sizer (OPS 3330, TSI, Minnesota, U.S.). Using the NanoScan SMPS (electrical equivalent mobility diameter: 0.01–0.42 µm, 13 channels at dry conditions) and the OPS (optical diameter: 0.3–10 µm, 13 channels at dry conditions) can provide a useful setup to monitor a wide particle diameter range 0.01–10 µm. However, combining the measurement results of these two instruments is challenging as will be pointed out in the next section.

The NanoScan SMPS consists of four main built-in components: (1) a cyclone inlet to remove large particles, (2) unipolar particle charger, (3) a radial differential mobility analyzer (RDMA), and (4) an isopropanol-based condensation particle counter (CPC). The particle number size distribution scan was 60 s (45 s upscan and 15 s downscan). The inlet flow rate was 0.75 lpm (±20%) whereas the sample flow rate was 0.25 lpm (±10%).

The OPS measured the particle number size distribution using the TSI default particle size bins, which consisted of 13 equally sized bins based on a lognormal scale. The dead-time correction was applied in the OPS operation. Sampling time-resolution was 5 min with a flow rate ~1 lpm.

The total number concentration of submicron aerosols was measured with a portable condensation particle counter (CPC 3007-2, TSI, Minnesota, U.S.). The cutoff size of this CPC was 10 nm and it was capable of measuring submicron particle number concentration of aerosols with diameters up to 2 µm. According to the specifications provided by the manufacturer, the maximum detectable concentration was 10^5 cm^{-3} with 20% accuracy. The sampling flow rate was 0.1 lpm (inlet flow rate 0.7 lpm).

Each instrument had its own aerosol sampling inlet (~1-m-long and 8 mm inner diameter) which was led through the wall to sample the outdoor aerosols. Each inlet consisted of short Tygon tubes (4 mm inner diameter) connected to a diffusion drier (TSI model 3062-NC). The diffusion drier was used to remove the excess moisture from the aerosol sample.

The aerosol transport efficiency through the aerosol inlet assembly was determined experimentally: ambient aerosol sampling alternatively with and without inlet. The aerosol data was corrected accordingly (Figure S2). The penetration efficiency was ~47% for 10 nm, ~93% for 0.3 μm, and ~40% for 10 μm particles. Accordingly, the particle number size distributions were corrected for losses in the tubing and the diffusion drier.

Based on the hourly mean total number concentrations, the ratio between SMPS (10–420 nm) and the CPC was approximately 0.86 and the relationship between them was linear with an R^2 = 0.78 (Figure S3a). The differences can arise from the fact that the CPC measured the full range of the submicron particles (from 10 nm up to a couple of micrometers) whereas the SMPS measurement range was nominally within the diameter range 10–420 nm. In practice, the NanoScan was not able to measure aerosols with diameters larger than 250 nm, which makes it an ultrafine ($D_p < 0.1$ μm) particle sizer. On the other hand, considering the number concentration within the diameter range 10–1000 nm, which can be obtained by the combined distributions measured with the SMPS and the OPS, revealed that its mean ratio compared to that measured with the CPC was about 1.14 and the relationship between them was linear with an R^2 = 0.78 (Figure S3b). This is expected because the CPC data was not corrected for sampling line losses.

2.2. Data Handling

The measurement time-resolution of the SMPS and the CPC was 1 min and of the OPS was 5 min. In order to construct a wide range of the measured particle number size distribution, we performed the following steps: (1) calculated the 5-min average of the SMPS data, (2) omitted the last two channels in the SMPS (i.e., remaining diameter range was 0.01–0.25 μm), (3) omitted the first channel in the OPS (i.e., remaining diameter range was 0.32–10 μm), and (4) merged the two distributions. As such, we obtained a combined particle number size distribution covering the diameter range 0.01–10 μm.

We calculated the particle number concentration (cm^{-3}) within four particle diameter ranges (size-fractionated number concentration): 0.01–0.025 μm (nucleation), 0.025–0.1 μm (Aitken), 0.1–1 μm (accumulation), and 1–10 μm (coarse). Consequently, the total number concentration was obtained as the sum of all these fractions. The size-fractionated number concentrations were obtained by integrating (practically summation) the measured particle number size distribution over the specified particle diameter range

$$PN_{D_{P2}-D_{P1}} = \int_{D_{P1}}^{D_{P2}} n_N^0 d\log_{10}(D_p), \quad (1)$$

where $n_N{}^0 = dN/d\log_{10}(D_p)$ is the measured particle number size distribution and D_p is the particle diameter.

The processed aerosol data (including all size-fractionated data) was then converted to hourly statistical analysis. This hourly averaged data was then used to calculate the daily and monthly statistical values. The statistical analysis included average, standard deviation, median, minimum, maximum, and percentiles (5%, 25%, 75%, and 95%) of valid number of data points, and percentage of valid data points.

2.3. Multi-Lognormal Distribution

A particle number size distribution can be characterized with a multi-lognormal distribution function [22]

$$\sum_{i=1}^{n} \frac{N_i}{\sqrt{2\pi}\log(\sigma_{g,i})} \exp\left[-\frac{\left(\log(D_p) - \log(D_{pg,i})\right)^2}{2\log^2(\sigma_{g,i})}\right], \quad (2)$$

where n is the number of individual log-normal modes and D_p is the aerosol particle diameter, and the three log-normal parameters that characterize an individual log-normal mode i are the mode number concentration N_i, the geometric variance $\sigma^2{}_{g,i}$, and the geometric mean diameter $D_{pg,i}$.

Here we used the DO-FIT algorithm to fit the particle number size distribution. The original DO-FIT algorithm was developed for the submicron particle size fraction [22]; here it was modified to include the coarse mode fraction. We ran the DO-FIT algorithm to fit the particle number size distributions (0.01–10 µm, 5-min resolution) by using a fixed number of modes, which was four.

2.4. Weather Conditions

The weather conditions were obtained from the continuous measurement performed on the rooftop of the Department of Physics. In this study, we selected weather data for the same time period as the aerosol measurement (i.e., 1 August 2016–31 July 2017; Figure 1).

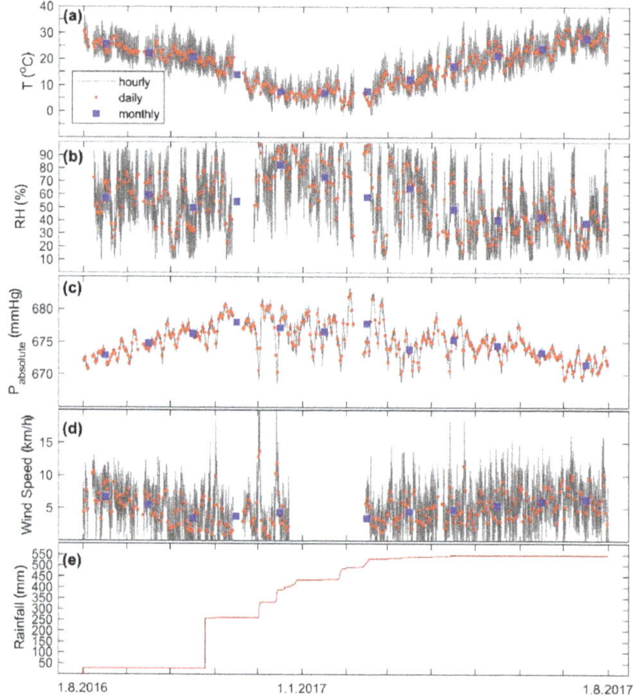

Figure 1. Time series of weather conditions during the measurement period (1 August 2016–31 July 2017) presented as hourly, daily, and monthly means for (**a**) ambient temperature, (**b**) relative humidity, (**c**) absolute pressure, and (**d**) wind speed. (**e**) The rainfall is presented as hourly cumulative precipitation.

The weather measurement was performed with a weather station (WH-1080, Clas Ohlson: Art.no. 36-3242). The time resolution of the measurement was 5 min. The weather data consisted of ambient temperature (−40–65 °C, resolution 0.1 °C), absolute pressure (918.7–1079.9 hPa, resolution 0.3 hPa), relative humidity (10%–99%, resolution 1%), wind speed (1–160 km/h) and direction (divided into 16 wind sectors), and precipitation (0–9999 mm, resolution 0.3 mm below 1000 mm and 1 mm over 1000 mm).

According to the daily mean (Figure 1), the temperature was 1–32 °C (overall mean 18 ± 8 °C) and the absolute pressure was 670–682 mmHg (overall mean 675 ± 3 mmHg). The overall average relative humidity was 53 ± 21% and the mean wind speed was 5 ± 3 km/h. The cumulative rain was about 550 mm.

3. Results and Discussion

3.1. Temporal Variation of the Particle Number Concentrations

The submicron particle number concentration (PN_{Sub}) was higher during the winter than during the summer (Figure 2a), suggesting a seasonal variation of fine aerosol number concentrations. Based on the monthly means, PN_{Sub} was in the range 3.3×10^4–3.7×10^4 cm^{-3} during the winter (December–February) and 1.2×10^4–1.6×10^4 cm^{-3} during the summer and early autumn (June–September) (Table S1). According to the 24-h mean number concentrations, PN_{Sub} was as high as 6.5×10^4 cm^{-3} and as low as 7.7×10^3 cm^{-3}. According to the hourly means, it was in the range 2.2×10^3–2.1×10^5 cm^{-3}. This seasonal variation was also confirmed by the CPC observations (Figure 2b).

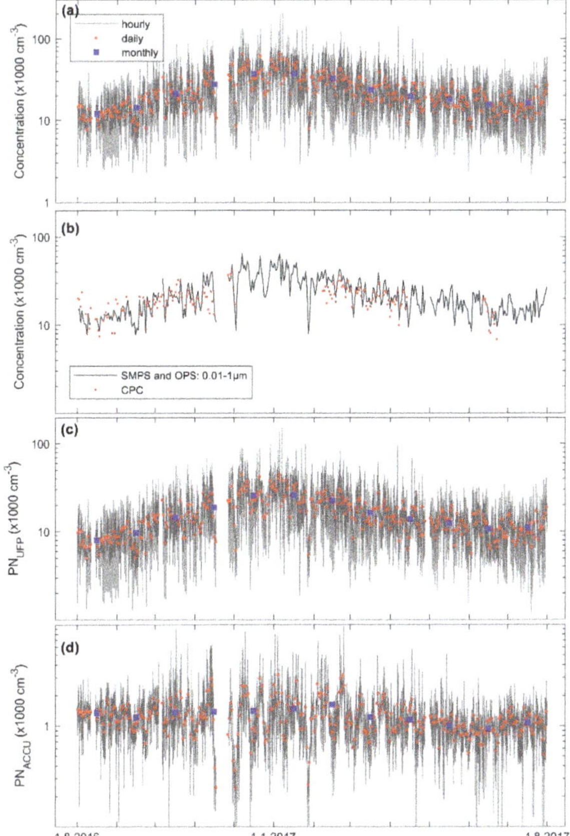

Figure 2. Time series of the (**a**) submicron particle number concentration, (**b**) comparison between the condensation particle counter (CPC) and scanning mobility particle sizer (SMPS) + optical particle sizer (OPS) particle number concentrations, and (**c**,**d**) the main particle size fraction concentrations of ultrafine particles (D_p < 0.1 μm) and accumulation mode particles (D_p 0.1–1 μm).

Looking closely at the hourly mean PN_{Sub} concentrations, we can clearly see a diurnal pattern that was more pronounced during the winter than during the summer (Figure 3). Basically, during the wintertime, the boundary layer is typically shallower and the aerosol emissions will be more concentrated in the lower atmosphere and their diurnal pattern becomes more pronounced. The main differences between the winter and summer diurnal patterns of the PN_{Sub} concentration were observed in the morning and late night, where the concentrations were high due to increased emissions from traffic during commuting hours. Similar to many other urban environments, the diurnal pattern observed in this study reflects the combustion emissions from traffic activity, which is more during the workdays [23,24]. These two peaks are relevant for the morning and afternoon traffic rush hours, which are similar to those noticed in most cities in the other countries [24–27].

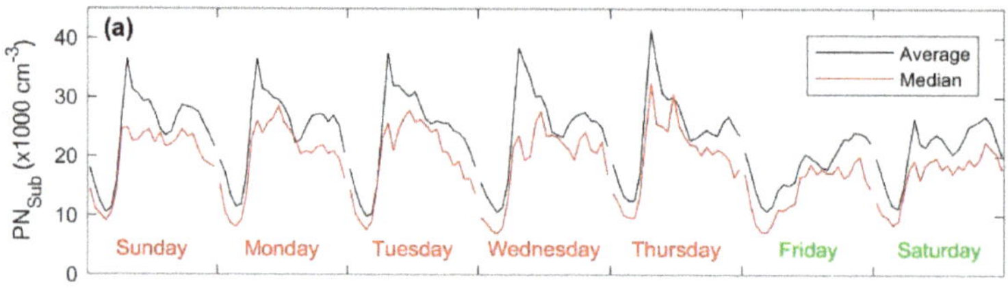

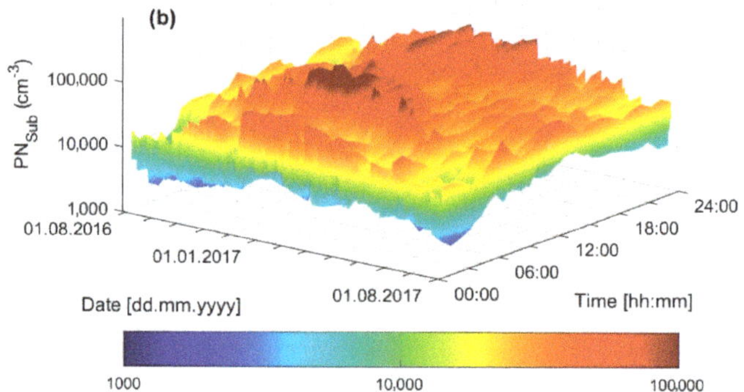

Figure 3. (**a**) Diurnal pattern of the submicron particle number concentration (0.01–1 μm) and (**b**) date-time spectrum showing the day-to-day and hour-to-hour variation of the number concentration; the color bar scales the number concentration (cm^{-3}).

Further analysis on the hourly mean PN_{Sub} concentration revealed a clear weekly cycle in the diurnal patterns with higher concentrations during the workdays (Sunday–Thursday) and lower concentrations during the weekend (Friday–Saturday); see Figure 3a. For example, during the weekends, the peak hourly mean PN_{Sub} concentration was below 2.7×10^4 cm^{-3} while during the workdays it was higher than 3.3×10^4 cm^{-3}. On workdays, PN_{Sub} diurnal pattern was characterized by a high concentration peak (as high as 4.1×10^4 cm^{-3}) during the morning traffic rush hours and another peak (as high as 2.9×10^4 cm^{-3}) during the afternoon. Regardless of the weekday, the lowest PN_{Sub} was below 1.5×10^4 cm^{-3} (on average as low as 1×10^4 cm^{-3}) and it was observed before the morning, when the traffic activity was at minimum in the city.

3.2. Concentrations of Different Particle Size-Fractions

The ultrafine particle number concentration (PN_{UFP}, diameter < 0.1 μm) fraction was about 93% of PN_{Sub} concentration. The accumulation mode particle number concentration (PN_{ACCU}, diameter 0.1–1 μm) was about 7% of PN_{Sub}. Table S1 lists the monthly statistical values for the main particle size fractions. Typically, the urban/suburban number concentration of fine particles is dominated by ultrafine particles [22,24]. This fact has been confirmed in several studies, both experimentally and theoretically [24–35].

In general, the temporal variation of PN_{UFP} concentration (Figures 2c, 4a and 5a) was very close to that of PN_{Sub} concentration; this is true for the seasonal variation, diurnal pattern, and the weekly cycle (Figures 2a and 3a,b). As for the PN_{ACCU} concentrations, the weekly cycle also showed distinguished diurnal patterns (Figures 2d and 4b). On workdays, the diurnal pattern was characterized by two main peaks, which had rather similar concentrations as high as 1.8×10^4 cm^{-3}. During the weekends (Friday–Saturday) and Thursday, the PN_{ACCU} concentration was exceptionally high during the late evening, possibly due to increased traffic activities during leisure time in the city.

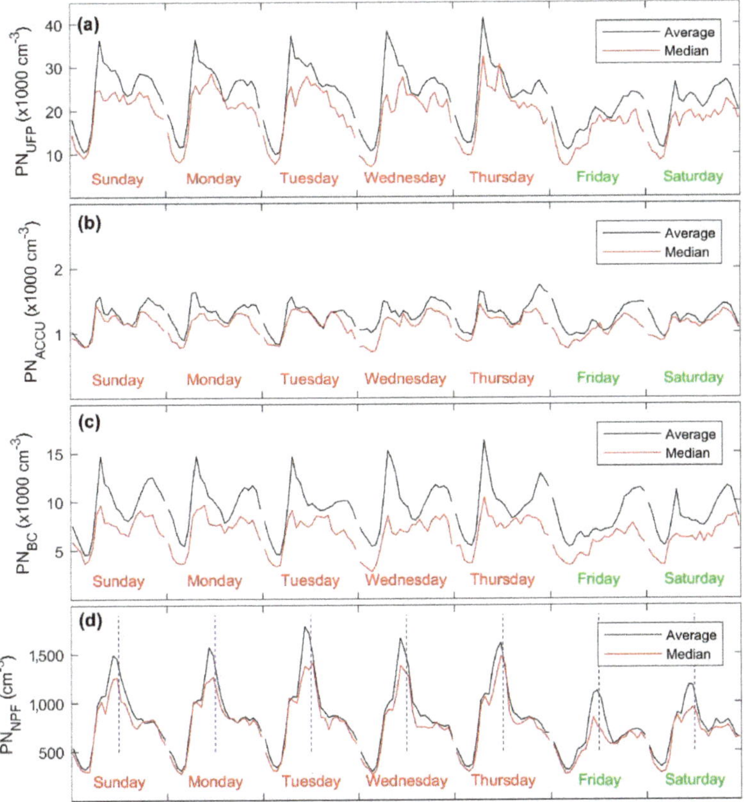

Figure 4. Diurnal pattern of the (**a**) ultrafine particle number concentration (UFP, diameter <0.1 µm), (**b**) accumulation mode particle number concentration (diameter 0.1–1 µm), (**c**) an indicator for the temporal variation of black carbon fraction (diameter 0.03–0.25 nm), and (**d**) number concentration in the first channel (diameter 10–15 nm) measured with SMPS; the vertical dashed lines are aligned with the noon time.

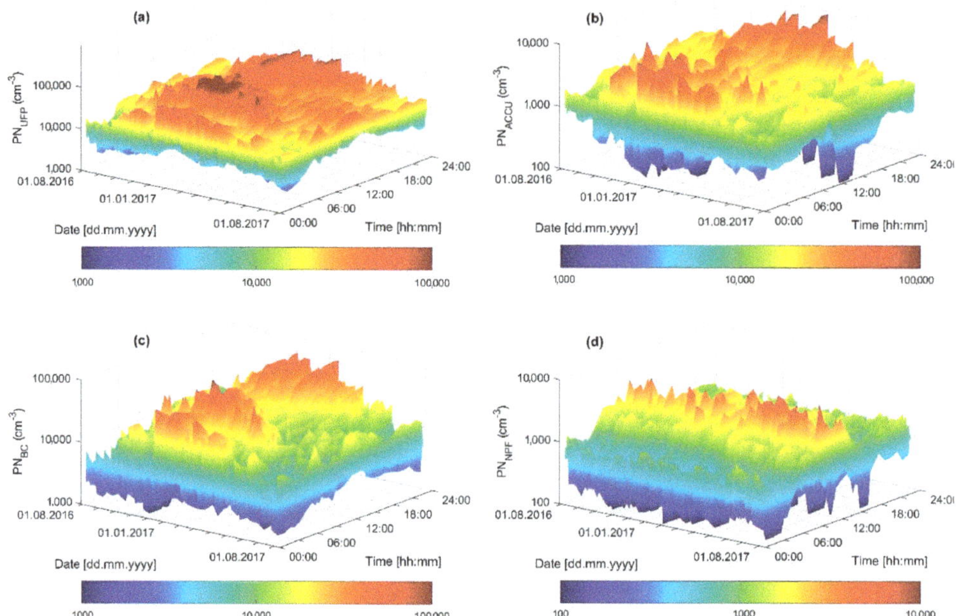

Figure 5. Date-time spectra showing the day-to-day and hour-to-hour variation of (**a**) ultrafine particle number concentration (UFP, diameter <0.1 µm), (**b**) accumulation mode particle number concentration (diameter 0.1–1 µm), (**c**) an indicator for the temporal variation of black carbon fraction (diameter 0.03–0.25 nm), (**d**) number concentration in the first channel (diameter 10–15 nm) measured with SMPS. The color bar scales the number concentration (cm^{-3}).

The coarse mode particle number (PN_{Coarse}) concentration had a different seasonal pattern than that of PN_{Sub} (Figure 6). According to the monthly average, the PN_{Coarse} was the highest during the autumn and spring (Table S1). For instance, the monthly PN_{Coarse} was about 2.5 and 3.8 cm^{-3}, respectively, in October and November. During March and April, the monthly PN_{Coarse} was about 3.3 cm^{-3}. As previously reported about this particle size fraction in Amman [19], it had a clear seasonal pattern linked to the sand and dust storms, which are often occurring in the spring season, and local dust resuspension, which are usually dominant in the autumn season. In general, the dust episodes lasted from a few hours to several days. Based on the daily average PN_{Coarse}, it was noticed that dust episodes occurred with concentrations that often exceeded 2 cm^{-3} and concentrations as high as 14.5 cm^{-3}. Based on the hourly average, the dust episodes increased PN_{Coarse} to values as high as 46 cm^{-3}. It is already well known that the black carbon has a well-defined size distribution in the submicron fraction. For instance, the distribution of black carbon in the ambient atmosphere, at road side, and from diesel emissions was found to span over the particle diameter range from around 30 nm up to a couple of hundred nanometers [36,37]. Here, we calculated the number concentrations within the particle diameter range 30–250 nm, which is an indicator of temporal variation of atmospheric black carbon (PN_{BC}). As expected, this size fraction had temporal variations (Figure 4c) very close to those of the total number concentrations but with lower concentrations (Figure 3a). It is important to mention here that what we presented here as PN_{BC} should not be understood as the concentration of black carbon; we only wanted to have an insight into the temporal variation of the black carbon fraction.

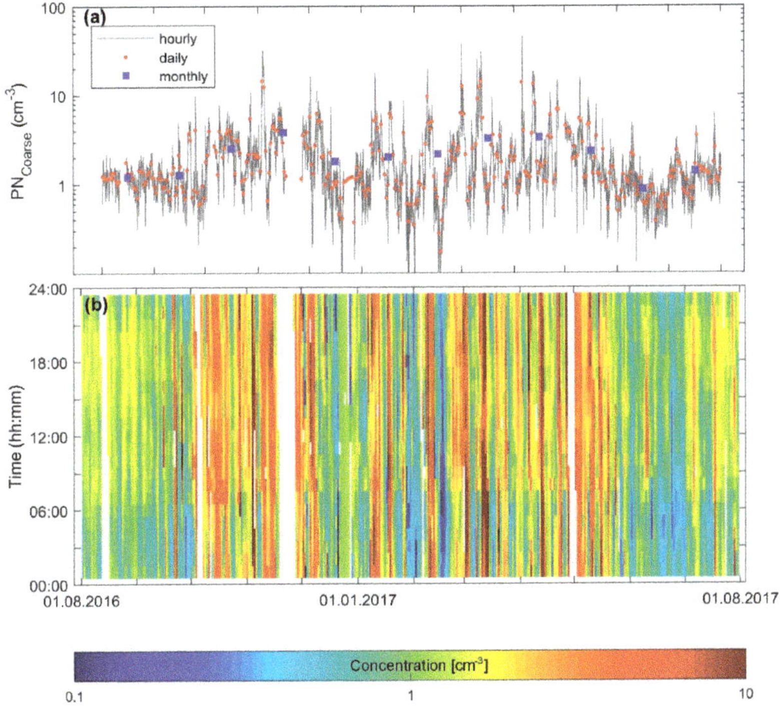

Figure 6. (**a**) Time series of the coarse mode (diameter 1–10 μm) particle number concentration and (**b**) date-time spectrum showing the day-to-day and hour-to-hour variation of the number concentration; the color bar scales the number concentration (cm^{-3}).

Except for Amman [17–21,38–40], particle number concentrations and size distributions has never been reported before in urban areas in the Middle East. The previous studies were mainly about some gaseous pollutants; PM mass concentrations, and chemical composition [41–56]. However, the submicron particle number concentrations in Amman are less than what was reported in megacities but still higher (sometimes comparable) than what was reported in Europe, US, and other parts in the world as will be shown in the next paragraph.

Comparison to observations in other urban locations is not straightforward. However, we can summarize the concentrations levels reported worldwide. Many studies reported number concentration in many urban and suburban environments in the EU, America, Asia, and Oceania [34,57–73]. The concentrations varied between cities, locations within the city, and time of observation reflecting the complex behavior of urban aerosols. The observed concentrations in this study (range 6.5×10^3–7.7×10^4 cm^{-3}) were comparable to other urban locations in the EU, the US, and Oceania where the reported concentrations were in the range 1×10^3–6×10^4 cm^{-3} [34,57–59,67–78]. We should keep in in mind that concentrations on roads are expected to be significantly higher than what can be observed in the ambient conditions [60–66]. Megacities exhibit the highest concentrations of air pollution. Here, we recall the results reported for Indian, Chinese, and Colombian cities [79–83]. Apte et al. [80] reported mean UFP concentrations in New Delhi (India) of 2.8×10^5 cm^{-3}, which was about eight times higher than that observed in ambient conditions in India. In Beijing and Xi'an (China), the concentrations are comparable to what can be observed in Delhi [81,82]. In Bogota (Colombia), Betancourt et al. [83] showed that the average fine particle number concentrations was about 2×10^5 cm^{-3}.

Secondary particle formation, which is often referred to as new particle formation (NPF), is an important process affecting particle number concentrations in both remote and urban environments [6–8,84]. The secondary particles can have a role in haze formation and influence the urban boundary layer and heat island. The NPF process starts from oxidation of precursor vapors and their clustering in sub-3 nm size range [84], which is outside the size range of the instrumentation used in this study. Therefore, it was not possible to make a constructive conclusion about the characteristics of these secondary aerosols in Amman. However, the first particle channel (diameter 10–15 nm) in the SMPS can be used as an indicator for the concentrations during NPF events, here denoted as PN_{NPF}.

The average diurnal pattern of PN_{NPF} was characterized by high concentrations (0.7×10^3–1.1×10^3 cm^{-3}) during the daytime and a sharp peak around midday (Figure 4d). The lowest concentration of this particle size fraction was ~320 cm^{-3} and it was observed between midnight and morning; specifically, between 3 a.m. and 4 a.m.

The most interesting part of the diurnal pattern in this size fraction is the sharp peak around midday. This peak was observed on ~34% of the measurement days (in total 110 days out of 326 days of valid data) and its alternating occurrence can be seen in Figure 5d. The peak was often observed around 11 a.m. with a concentration in the range 1.5×10^3–1.8×10^3 cm^{-3} on workdays and ~1.1×10^3 cm^{-3} on weekend days. The concentration at the edge of the size distribution started to increase around 8 a.m. and the increase was over around 1 p.m. New particle formation characterization requires deep analysis, which is beyond the scope of this study; therefore, we will not explain this phenomenon any further here.

3.3. Modal Structure of the Particle Number Size Distribution

For the purposes of investigating the temporal variation of the particle number concentrations, we focused on two periods representing cold (December–February) and warm (May–August) conditions separately (Figures S4 and S5). During the cold period, the weekly cycle of the diurnal pattern was very distinguished with high mean concentrations on workdays and low mean concentrations on weekends (Figure S4 and Figure 7a–f). On all weekdays, the diurnal pattern was characterized by two peaks: morning and afternoon. The afternoon peak (highest concentration was in the range 3×10^4–3.5×10^4 cm^{-3}) was rather similar on all weekdays; however, the first peak was higher on workdays than on weekend days. Furthermore, the first peak tended to have higher concentrations from Monday (~4.5×10^4 cm^{-3}) through Thursday (~6.5×10^4 cm^{-3}). During this cold period and on all weekdays, the mean particle number size distributions during daytime hours had their peak diameter below 40 nm whereas during the evening it was over 40 nm (Figure S6a,b).

During the warm period, the diurnal pattern was rather similar on all weekdays (Figure S5). The daytime mean total concentration (1×10^4–2×10^4 cm^{-3}) during this warm period was lower than that during the cold period (often higher than 3×10^4 cm^{-3}). The differences are possibly explained by the ambient conditions such as enhanced wind speed, higher temperature, and higher boundary layer during the warm period. The secondary particle formation (i.e., NPF events) was very pronounced during the warm period. NPF events were observed almost every day and, consequently, they can be seen to dominate the mean particle number size distribution (Figure 7g–l). The newly formed aerosols were clearly observed with their distinguished nucleation mode with high concentrations before the noon (specifically between 9 a.m. and noon; Figure 7i,j). These newly formed aerosols displayed growth until 6 p.m., when their geometric mean diameter (GMD) reached ~40 nm.

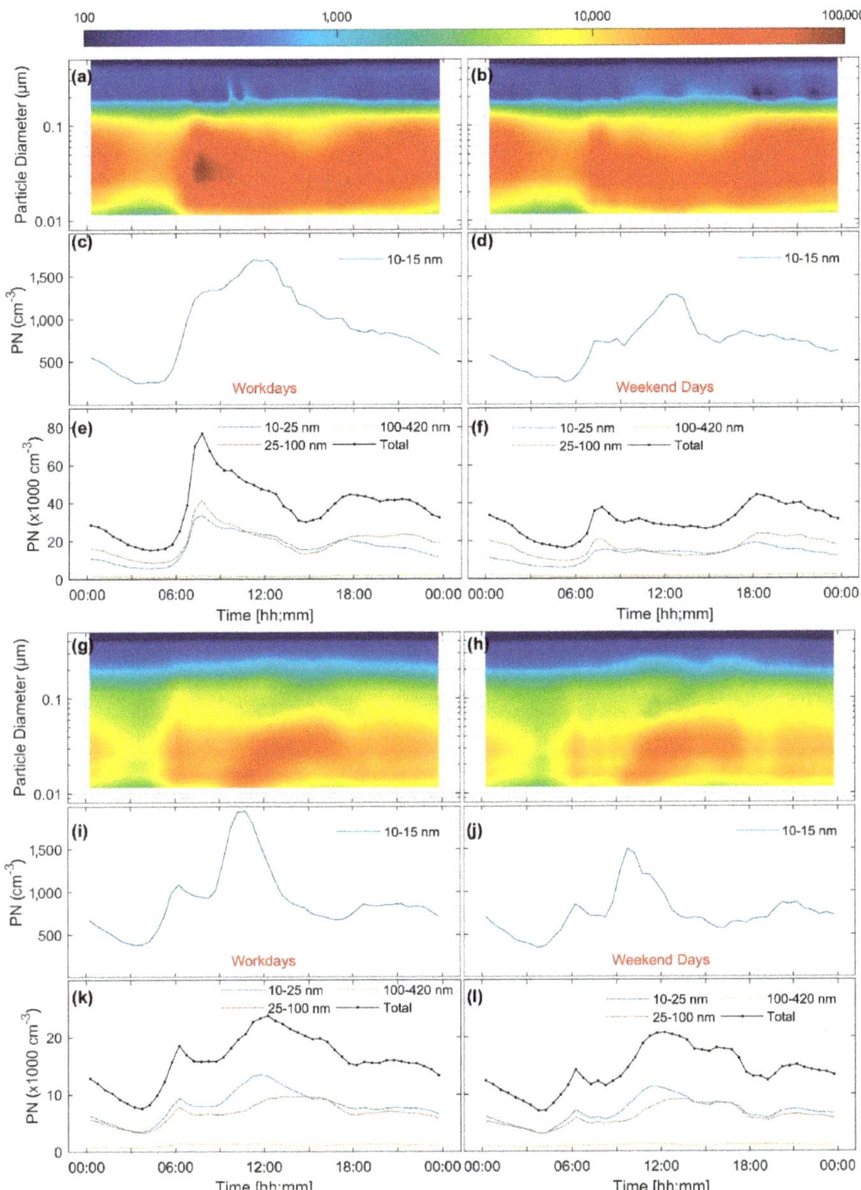

Figure 7. Mean diurnal patterns during (**a–f**) a cold period (i.e., December–February) and (**g–l**) warm period (i.e., May–August). Here the color subfigures (**a**,**b**,**g**,**h**) show the mean particle number size distribution spectra (color bar represents dN/dlog(Dp) (cm^{-3})) and below them are plots for different particle size fractions.

Contrary to the case during the cold period, the mean particle number size distributions at different times of the day during the warm period had three well-separated submicron modes: nucleation, Aitken, and accumulation, in addition to a mode with small GMD that represented the early stage of NPF events (Figure S6c,d).

Further analysis of the particle number size distributions with respect to the lognormal fitting (i.e., modal structure) revealed that the mode geometric mean diameter ($D_{pg,i}$) was about 22 nm, 62 nm, 225 nm, and 2.23 μm for the nucleation mode, Aitken mode, accumulation mode, and coarse mode, respectively (Figure 8a,b). The corresponding geometric mean values of the mode number concentrations (N_i) were about 9.4 × 10^3, 3.9 × 10^3, 158, and 1.2 cm^{-3}, respectively (Figure 8c–f). The mode number concentrations of the nucleation, Aitken, and coarse modes were lognormally distributed around their geometric mean values: 22 nm, 62 nm, and 2.3 μm; respectively. However, the accumulation mode number concentration had two distinguished modes with concentrations centered around 50 cm^{-3} (centered around D_{pg} = 390 nm) and 790 cm^{-3} (centered around D_{pg} = 165 nm); see also Figure 8a. As for the mode geometric variance ($\sigma_{pg,i}$), the geometric mean value was about 1.73, 1.67, 1.54, and 1.62, respectively for the nucleation mode, Aitken mode, accumulation mode, and coarse mode (Figure 8g–j). Interestingly, and as can be seen from Figure 8a, D_{pg} of the nucleation mode reached values close to 2 nm, which is well below the cut-off diameter of the SMPS (10 nm). That was basically due to the occurrence of NPF events when a part of the nucleation mode was below the cut-off diameter of the SMPS, which was 10 nm. Furthermore, the smaller the D_{pg}, the higher the N_i was for the nucleation mode when it occurred below 10 nm.

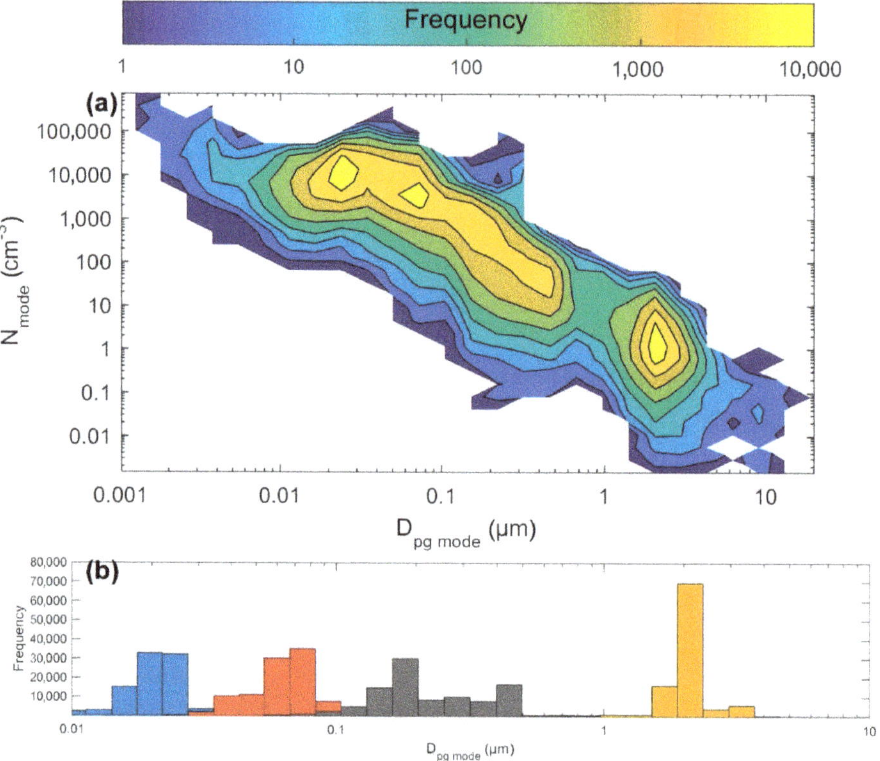

Figure 8. *Cont.*

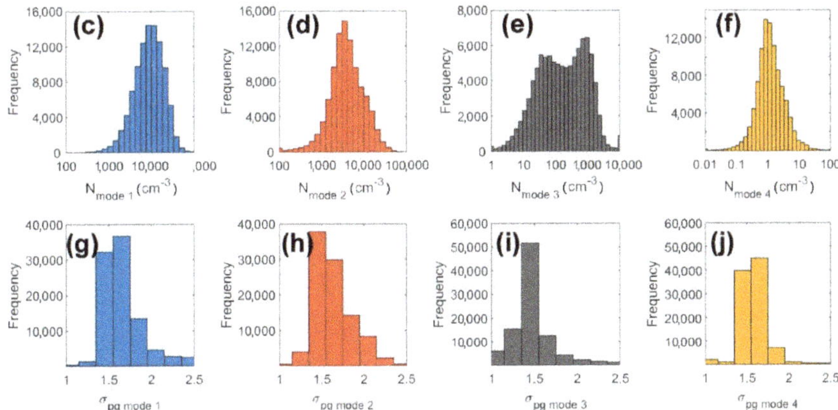

Figure 8. Modal structure of the particle number size distribution derived from the multi-lognormal fitting (5-min average). (**a**) Modal structure spectrum showing the dominant modes (nucleation, Aitken, accumulation, and coarse modes). Distributions of (**b**) mode geometric mean diameter (D_{pg}), (**c–f**) mode number concentration (N), and (**g–j**) mode geometric variance (σ_{pg}).

3.4. Effect of Local Weather Conditions

The meteorological conditions affected the aerosol concentrations in different ways. For example, the submicron particles were affected in a different way than coarse mode particles. Understanding the relationship between urban aerosol concentrations and local meteorological conditions makes it possible to develop predictive models for air pollution [85–88].

In general, the wind speed had a major influence on the particle number concentration (Figure 9). The PN_{Sub} had a decreasing trend with respect to the wind speed (Figure 9a). For instance, the mean PN_{Sub} was about 2.8 × 10^4 cm^{-3} at stagnant conditions and it reached below 1.2 × 10^4 cm^{-3} when the wind speed was above 20 km/h. This indicates that most of the submicron fraction originated from local sources such as combustion processes (e.g., traffic tailpipe emissions). However, during wind speed 10–15 km/h, the PN_{Sub} slightly increased to about 1.2 × 10^4 cm^{-3} and then decreased again with increasing wind speed.

The PN_{Coarse} concentrations had a U-shaped (centered around 7.5 km/h at about 1.5 cm^{-3}) with respect the wind speed (Figure 9b). Below 7.5 km/h, the PN_{Coarse} concentrations were increasing (reaching about 2.1 cm^{-3} at stagnant conditions) with decreasing wind speed and above that wind speed the PN_{Coarse} were increasing (reaching about 5.5 cm^{-3} at wind speed 20 km/h) with increasing wind speed. The left-hand side of the U-shape curve is related to local sources of dust not induced by the wind; i.e., road dust resuspension due to traffic activities. The right-hand side of the U-shape curve is related to local dust resuspension induced by the wind. Another reason for the increased PN_{Coarse} concentration at high wind speed is the sand and dust storm (SDS) via long-range transport. The observed influence of wind speed on PN_{Sub} and PN_{Coarse} concentrations in this study is consistent with previous observations in other urban environments [26,87–95].

The investigation of the effect of temperature on the aerosol concentrations is very complex as illustrated in Figure 9c,d. The mean PN_{Sub} had the highest concentrations (~3 × 10^4 cm^{-3}) during cold conditions (T < 10 °C) and the lowest concentrations (~1.5 × 10^4 cm^{-3}) during moderate temperature conditions (T in the range 15–25 °C). The mean PN_{Sub} concentrations was proportional to the temperature, when it increased from 25 to 40 °C. The PN_{Coarse} concentrations were proportional to the temperature in the range −5–10 °C and it was rather constant (1–2 cm^{-3}) when the temperature was higher than 10 °C. Our results seem to be in consistent with [96], who found that aerosol concentrations increased with relative humidity during the winter season and decreased with temperature. Additionally, Olivares et al. [26] reported that the increase in particle number concentration with decreasing temperature is different for different particle size, and they also showed a distinct correlation between number concentration and (temperature and relative humidity) with higher concentrations during periods with low temperatures or high relative humidity.

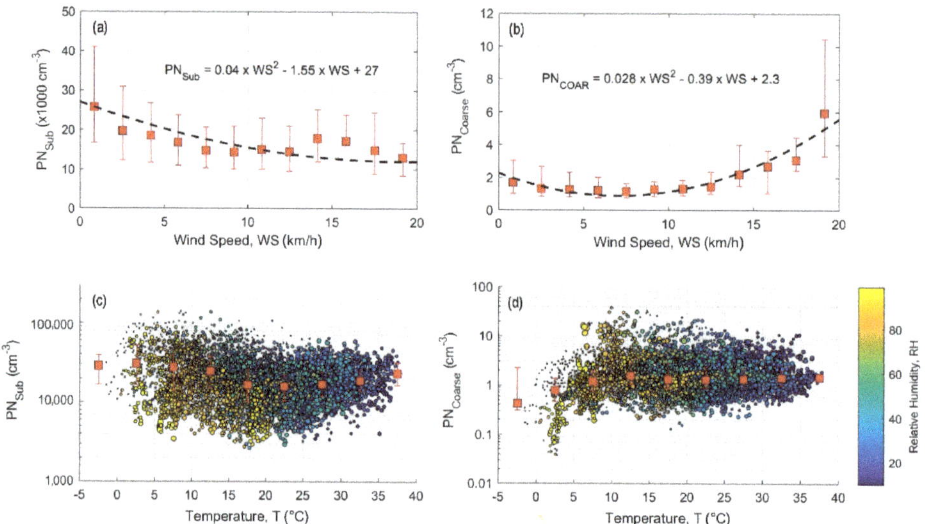

Figure 9. Particle number concentration dependence on the local weather conditions: (**a**) submicron particles (diameter < 1 µm) with respect to wind speed, (**b**) coarse mode particles (1–10 µm) with respect to wind speed, (**c**) submicron particles with respect to temperature, and (**d**) coarse mode particles with respect to temperature. The red squares are the median concentrations with respect to temperature intervals on the x-axis and the error bars are the 25% and 75%. In (**c**,**d**), the color codes the relative humidity and the circle size is proportional to the wind speed.

4. Conclusions

In this study we aimed at characterizing the aerosol particle number size distribution. We explored the temporal variability (diurnal, weekly, and seasonal) of the number concentrations in Amman, Jordan. Amman can be considered as an example for Middle Eastern urban conditions. Here we utilized a long-term measurement of particle number size distributions over a wide particle diameter range (0.01–10 µm), which was combined from a Nano-Scanning Mobility Particle Sizer (Nano SMPS) and an Optical Particle Sizer (OPS). We also investigated the modal structure of the aerosol population by using the multi-lognormal distribution function (DO-FIT).

The submicron particle number concentrations (PN_{Sub}) and the coarse mode particle number concentration (PN_{Coarse}) had distinguished seasonal variations. Based on the monthly average, PN_{Sub} was higher in the winter (3.3×10^4–3.7×10^4 cm^{-3}, December–February) than in the summer and early autumn (1.2×10^4–1.6×10^4 cm^{-3}, June–September). The 24-h mean PN_{Sub} was in the range 6.5×10^3–7.7×10^4 cm^{-3} and the hourly means was in the range 2.2×10^3–2.1×10^5 cm^{-3}. According to the monthly average, the PN_{Coarse} was the highest during the autumn (3.8 cm^{-3}, November) and spring (3.3 cm^{-3}, April). The PN_{Coarse} seasonal pattern is most likely linked to the sand and dust storms (often occurring in spring) and local dust resuspension (dominant autumn). It was clear within this study that there is a clear influence of new particle formation (NPF) on the submicron aerosols and Sand and Dust Storm (SDS) on coarse mode aerosols. However, these phenomena require further detailed analysis that will be made in upcoming studies.

The PN_{Sub} concentrations was characterized by a diurnal pattern, which was more pronounced during the winter than during the summer. The diurnal pattern was also different between workdays and weekends reflecting a clear weekly cycle: higher concentrations during the workdays (Sunday–Thursday; over 3.3×10^4 cm^{-3}) and lower concentrations during the weekend (Friday–Saturday; below 2.7×10^4 cm^{-3}). On workdays, PN_{Sub} diurnal pattern was characterized with a high concentration peak (as high as 4.1×10^4 cm^{-3}) during the morning traffic rush hours and another peak (as high as 2.9×10^4 cm^{-3}) during the afternoon. In general, the seasonal, weekly, and diurnal variation of PN_{Sub} is clearly depicting the traffic activities suggesting that traffic emissions are a major contributor to submicron aerosols. The ultrafine particle number concentration (PN_{UFP}, diameter < 0.1 µm) fraction was about 93% of PN_{Sub} concentration whereas that of the accumulation mode particle number concentration (PN_{ACCU}, diameter 0.1–1 µm) was about 7% of PN_{Sub}. The temporal variation of PN_{UFP} and PN_{ACCU} concentration was very close to that of PN_{Sub} concentration.

We investigated the modal structure of the measured aerosols (0.01–10 mm). During cold conditions (December–February), the mean particle number size distributions during daytime hours had peak diameters smaller than 40 nm whereas during the evening it was larger than 40 nm. During warm conditions (May–August), the mean particle number size distributions at different times of the day had three well-separated submicron modes: nucleation, Aitken, and accumulation in addition to a mode with small GMD that represented the early stage of NPF events. The mode geometric mean diameter ($D_{pg,i}$) was about 22 nm, 62 nm, 225 nm, and 2.23 µm for the nucleation mode, Aitken mode, accumulation mode, and coarse mode, respectively. The corresponding mode number concentrations (N_i) were about 9.4×10^3, 3.9×10^3, 158, and 1.2 cm^{-3}, respectively. In fact, the accumulation mode number concentration had two distinguished groups with concentrations centered around 50 cm^{-3} (centered around D_{pg} = 165 nm) and 790 cm^{-3} (centered around D_{pg} = 390 nm). Interestingly, D_{pg} of the nucleation mode reached values close to 2 nm, which is well below the cut-off diameter of the SMPS (10 nm) but still can be depicted by the DO-FIT algorithm of the multi-lognormal fitting. That was basically due to the occurrence of NPF events.

The local meteorological conditions affected the aerosol concentrations in different ways. The wind speed and temperature had a major impact on the aerosol concentrations. During stagnant conditions, the mean PN_{Sub} concentration was ~2.8×10^4 cm^{-3} and PN_{Coarse} concentration was ~2.1 cm^{-3}. In general, the PN_{Coarse} concentration had a U-shape with respect to wind speed whereas the PN_{Sub} concentrations decreased with wind speed. The mean PN_{Sub} had the highest concentrations (~3×10^4 cm^{-3}) during cold conditions and the lowest concentrations (~1.5×10^4 cm^{-3}) during moderate temperature conditions. The PN_{Coarse} concentrations were proportional to the temperature in the range −5–10 °C and they were rather constant (1–2 cm^{-3}) when the temperature was higher than 10 °C.

Supplementary Materials: The following are available online at http://www.mdpi.com/2073-4433/10/11/710/s1, Figure S1: (a) A Map of Jordan showing the geographical location of Amman. (b) A Map of Amman showing the campus of the University of Jordan (shaded area) and (c) a detailed map of the campus of the University of Jordan, showing the sampling location (shaded area) in the middle of the campus., Figure S2: Experimental penetration efficiency through the sampling lines (tubing and diffusion drier), Figure S3: Regression between the number concentrations measured with the CPC versus (a) number concentration measured with the SMPS (*Dp* 0.01–0.42 µm) and (b) number concentration derived from the combined particle number size distribution measurements (SMPS and OPS, *Dp* 0.01–1 µm), Figure S4: Weekly cycle and the diurnal pattern during a cold period (i.e., winter: December–February): (a) mean particle number size distribution spectra and (b–c) concentrations of different particle size fractions. The color bar represents dN/dlog (*Dp*) (cm^{-3}), Figure S5: Weekly cycle and the diurnal pattern during a cold period (i.e., summer: May–August): (a) mean particle number size distribution spectra and (b–c) concentrations of different particle size fractions. The color bar represents dN/dlog (*Dp*) (cm^{-3}), Figure S6: Mean particle number size distributions during (a,b) cold period (i.e., winter: December–February) and (c,d) warm period (i.e., summer: May–August). The mean distributions are shown for different times of the day on workdays (a,c) and weekend days (b,d), Table S1: Monthly statistics for the main particle size fractions (concentrations in units of cm^{-3}).

Author Contributions: Conceptualization, T.H.; methodology, T.H.; validation, T.H.; formal analysis, T.H., S.H. and L.D.; investigation, T.H.; resources, T.H.; data curation, T.H.; writing—original draft preparation, T.H.; writing—review and editing, T.H., L.D., S.H., T.P. and M.K.; visualization, T.H. and L.D.; supervision, T.H.; project administration, T.H.; funding acquisition, T.H., T.P. and M.K.

Funding: This research was funded by the Scientific Research Support Fund (SRF, project number BAS-1-2-2015) at the Jordanian Ministry of Higher Education and the Deanship of Academic Research (DAR, project number 1516) at the University of Jordan. This research was part of a close collaboration between the University of Jordan and the Institute for Atmospheric and Earth System Research (INAR / Physics, University of Helsinki) via ERC advanced grant No. 742206, the European Union's Horizon 2020 research and innovation program under grant agreement No. 654109, the Academy of Finland Center of Excellence (project No. 272041), ERA-PLANET (www.era-planet.eu), trans-national project SMURBS (www.smurbs.eu, grant agreement n. 689443 funded under the EU Horizon 2020 Framework Programme and Academy of Finland via the Center of Excellence in Atmospheric sciences and NanoBioMass (project number 1307537).

Acknowledgments: This manuscript was written and completed during the sabbatical leave of the first author (Tareq Hussein) that was spent at the University of Helsinki and supported by the University of Jordan during 2019. Open access funding provided by University of Helsinki.

Conflicts of Interest: The authors declare no conflict of interest.

References

1. Lelieveld, J.; Evans, J.S.; Fnais, M.; Giannadaki, D.; Pozzer, A. The contribution of outdoor air pollution sources to premature mortality on a global scale. *Nature* **2015**, *525*, 367–371. [CrossRef] [PubMed]
2. IPCC. *Climate Change 2013: The Physical Science Basis. Contribution of Working Group I to the Fifth Assessment Report of the Intergovernmental Panel on Climate Change*; Cambridge University Press: Cambridge, MA, USA, 2013.
3. Oberdörster, G.; Sharp, Z.; Atudorei, V.; Elder, A.; Gelein, R.; Kreyling, W.; Cox, C. Translocation of inhaled ultrafine particles to the brain. *Inhal. Toxicol.* **2004**, *16*, 437–445. [CrossRef] [PubMed]
4. Kerminen, V.M.; Paramonov, M.; Anttila, T.; Riipinen, I.; Fountoukis, C.; Korhonen, H.; Asmi, E.; Laakso, L.; Lihavainen, H.; Swietlicki, E.; et al. Cloud condensation nuclei production associated with atmospheric nucleation: A synthesis based on existing literature and new results. *Atmos. Chem. Phys.* **2012**, *12*, 12037–12059. [CrossRef]
5. Merikanto, J.; Spracklen, D.; Mann, G.; Pickering, S.; Carslaw, K. Impact of nucleation on global CCN. *Atmos. Chem. Phys.* **2009**, *9*, 8601–8616. [CrossRef]
6. Kulmala, M.; Vehkamäki, H.; Petäjä, T.; Dal Maso, M.; Lauri, A.; Kerminen, V.-M.; Birmili, W.; McMurry, P.H. Formation and growth rates of ultrafine atmospheric particles: A review of observations. *J. Aerosol Sci.* **2004**, *35*, 143–176. [CrossRef]
7. Kerminen, V.-M.; Chen, X.; Vakkari, V.; Petäjä, T.; Kulmala, M.; Bianchi, F. Atmospheric new particle formation and growth: Review of field observations. *Environ. Res. Lett.* **2018**, *13*, 103003. [CrossRef]
8. Chu, B.; Kerminen, V.M.; Bianchi, F.; Yan, C.; Petäjä, T.; Kulmala, M. Atmospheric new particle formation in China. *Atmos. Chem. Phys.* **2019**, *19*, 115–138. [CrossRef]
9. Rönkkö, T.; Kuuluvainen, H.; Karjalainen, P.; Keskinen, J.; Hillamo, R.; Niemi, J.V.; Pirjola, L.; Timonen, H.J.; Saarikoski, S.; Saukko, E.; et al. Traffic is a major source of atmospheric nanocluster aerosol. *Proc. Natl. Acad. Sci. USA* **2017**, *114*, 7549–7554. [CrossRef]
10. Dall'Osto, M.; Harrison, R.M. Urban organic aerosols measured by single particle mass spectrometry in the megacity of London. *Atmos. Chem. Phys.* **2012**, *12*, 4127–4142. [CrossRef]
11. Pikridas, M.; Riipinen, I.; Hildebrandt, L.; Kostenidou, E.; Manninen, H.; Mihalopoulos, N.; Kalivitis, N.; Burkhart, J.F.; Stohl, A.; Kulmala, M.; et al. New particle formation at a remote site in the eastern Mediterranean. *J. Geophys. Res. Atmos.* **2012**, *117*. [CrossRef]
12. Ždímal, V.; Smolík, J.; Eleftheriadis, K.; Wagner, Z.; Housiadas, C.; Mihalopoulos, N.; Mikuška, P.; Večeřa, Z.; Kopanakis, I.; Lazaridis, M. Dynamics of atmospheric aerosol number size distributions in the eastern Mediterranean during the "SUB-AERO" Project. *Water Air Soil Pollut.* **2011**, *214*, 133–146. [CrossRef]
13. Reche, C.; Viana, M.; Pandolfi, M.; Alastuey, A.; Moreno, T.; Amato, F.; Ripoll, A.; Querol, X. Urban NH3 levels and sources in a Mediterranean environment. *Atmos. Environ.* **2012**, *57*, 153–164. [CrossRef]

14. Kopanakis, I.; Chatoutsidou, S.; Torseth, K.; Glytsos, T.; Lazaridis, M. Particle number size distribution in the eastern Mediterranean: Formation and growth rates of ultrafine airborne atmospheric particles. *Atmos. Environ.* **2013**, *77*, 790–802. [CrossRef]
15. Lihavainen, H.; Alghamdi, M.A.; Hyvärinen, A.-P.; Hussein, T.; Aaltonen, V.; Abdelmaksoud, A.S.; Al-Jeelani, H.; Almazroui, M.; Almehmadi, F.M.; Al Zawad, F.M.; et al. Aerosols Physical properties at Hada Al Sham, Western Saudi Arabia. *Atmos. Environ.* **2016**, *135*, 109–117. [CrossRef]
16. Hakala, S.; Alghamdi, M.A.; Paasonen, P.; Vakkari, V.; Khoder, M.; Neitola, K.; Dada, L.; Abdelmaksoud, A.S.; Al-Jeelani, H.; Shabbaj, I.I.; et al. New particle formation, growth and apparent shrinkage at a rural background site in western Saudi Arabia. *Atmos. Chem. Phys.* **2019**, *19*, 10537–10555. [CrossRef]
17. Hussein, T.; Boor, B.E.; dos Santos, V.N.; Kangasluoma, J.; Petäjä, T.; Lihavainen, H. Mobile Aerosol Measurement in the Eastern Mediterranean—A Utilization of Portable Instruments. *Aerosol Air Qual. Res.* **2017**, *17*, 1875–1886. [CrossRef]
18. Hussein, T.; Sogacheva, L.; Petäjä, T. Accumulation and Coarse Modes Particle Concentrations during Dew Formation and Precipitation. *Aerosol Air Qual. Res.* **2018**, *18*, 2929–2938. [CrossRef]
19. Hussein, T.; Juwhari, H.; Al Kuisi, M.; Alkattan, H.; Lahlouh, B.; Al-Hunaiti, A. Accumulation and Coarse Modes Aerosols Concentrations and Carbonaceous Contents in the Urban Background Atmosphere in Amman—Jordan. *Arab. J. Geosci.* **2018**, *11*, 617. [CrossRef]
20. Hussein, T.; Saleh, S.S.A.; dos Santos, V.N.; Abdullah, H.; Boor, B.E. Black Carbon and Particulate Matter Concentrations in Eastern Mediterranean Urban Conditions—An Assessment Based on Integrated Stationary and Mobile Observations. *Atmosphere* **2019**, *10*, 323. [CrossRef]
21. Hussein, T.; Betar, A. Size-Fractionated Number and Mass Concentrations in the Urban Background Atmosphere during Spring 2014 in Amman—Jordan. *Jordan J. Phys.* **2017**, *10*, 51–60.
22. Hussein, T.; Dal Maso, M.; Petäjä, T.; Koponen, I.K.; Paatero, P.; Aalto, P.P.; Hämeri, K.; Kulmala, M. Evaluation of an automatic algorithm for fitting the particle number size distributions. *Boreal Environ. Res.* **2005**, *10*, 337–355.
23. Hussein, T.; Hämeri, K.; Kulmala, M. Long-term indoor-outdoor aerosol measurement in Helsinki, Finland. *Boreal Environ Res.* **2002**, *7*, 141–150.
24. Hussein, T.; Puustinen, A.; Aalto, P.P.; Mäkelä, J.M.; Hämeri, K.; Kulmala, M. Urban aerosol number size distributions. *Atoms. Chem. Phys.* **2004**, *4*, 391–411. [CrossRef]
25. Wehner, B.; Wiedensohler, A. Long term measurements of submicrometer urban aerosols: Statistical analysis for correlations with meteorological conditions and trace Gases. *Atoms. Chem. Phys.* **2003**, *3*, 867–879. [CrossRef]
26. Olivares, G.; Johansson, C.; Ström, J.; Hansson, H.C. The role of ambient temperature of particle number concentrations in a street canyon. *Atmos. Environ.* **2007**, *41*, 2145–2166. [CrossRef]
27. Wehner, B.; Wiedensohler, A.; Tuch, T.M.; Wu, Z.J.; Hu, M.; Slanina, J.; Kiang, C.S. Variability of the Aerosol Number Size Distribution in Beijing, China: New Particle Formation, Dust Storms, and High continental Background. *Geophys. Res.* **2004**, *31*. [CrossRef]
28. Wehner, B.; Wiedensohler, A.; Heintzenberg, J. Submicrometer aerosol size distributions and mass concentration of the Millenium fireworks 2000 in Leipzig, Germany. *J. Aerosol Sci.* **2000**, *31*, 1489–1493. [CrossRef]
29. Jung, C.H.; Kim, Y.P.; Lee, K.W. Simulation of the influence of coarse mode particles on the properties of fine mode particles. *Aerosol Sci. Technol.* **2002**, *33*, 1201–1216. [CrossRef]
30. Mönkkönen, P.; Koponen, I.K.; Lehtinen, K.E.J.; Uma, R.; Srinivasan, D.; Hämeri, K.; Kulmala, M. Technical note: Death of nucleation and Aitken mode particles: Observations at extreme atmospheric conditions and their theoretical explanation. *J. Aerosol Sci.* **2004**, *35*, 781–787. [CrossRef]
31. Cheng, T.; Lu, D.; Chen, H.; Xu, Y. Physical characteristics of dust aerosol over Hunshan Dake sandland in Northern China. *Atmos. Environ.* **2005**, *39*, 1237–1243. [CrossRef]
32. Mejıa, J.F.; Morawska, L.; Mengersen, K. Spatial variation in particle number size distributions in a large metropolitan area. *Atmos. Chem. Phys.* **2008**, *8*, 1127–1138. [CrossRef]
33. Wu, Z.; Hu, M.; Lin, P.; Liu, S.; Wehner, B.; Widensohler, A. Particle number size distribution in the urban atmosphere of Beijing, China. *Atmos. Environ.* **2008**, *42*, 7967–7980. [CrossRef]

34. Birmili, W.; Alaviippola, B.; Hinneburg, D.; Knoth, O.; Touch, T.; Borken-Kleefeld, J.; Schacht, A. Dispersion of traffic-related exhaust particles near the Berlin urban motorway—Estimation of fleet emission factors. *Atmos. Chem. Phys.* **2009**, *9*, 2355–2374. [CrossRef]
35. Oliveira, C.; Alves, C.; Pio, C.A. Aerosol particle size distributions at a traffic exposed site and an urban background location in Oporto, Portugal. *Quim. Nova* **2009**, *32*, 928–933. [CrossRef]
36. Ning, Z.; Chan, K.L.; Wong, K.C.; Westerdahl, D.; Močnik, G.; Zhou, J.H.; Cheung, C.S. Black carbon mass size distributions of diesel exhaust and urban aerosols measured using differential mobility analyzer in tandem with Aethalometer. *Atmos. Environ.* **2013**, *80*, 31–40. [CrossRef]
37. Wu, Y.; Wang, X.; Tao, J.; Huang, R.; Tian, P.; Cao, J.; Zhang, L.; Ho, K.-F.; Han, Z.; Zhang, R. Size distribution and source of black carbon aerosol in urban Beijing during winter haze episodes. *Atmos. Chem. Phys.* **2017**, *17*, 7965–7975. [CrossRef]
38. Hussein, T.; Abu Al-Ruz, R.; Petäjä, T.; Junninen, H.; Arafah, D.-E.; Hämeri, K.; Kulmala, M. Local air pollution versus short–range transported dust episodes: A comparative study for submicron particle number concentration. *Aerosol Air Qual. Res.* **2011**, *11*, 109–119. [CrossRef]
39. Hussein, T.; Halayka, M.; Abu Al-Ruz, R.; Abdullah, H.; Mølgaard, B.; Petäjä, T. Fine Particle Number Concentrations in Amman and Zarqa during Spring 2014. *Jordan J. Phys.* **2016**, *9*, 31–46.
40. Saleh, S.S.A.; Shilbayeh, Z.; Alkattan, H.; Al-Refie, M.R.; Jaghbeir, O.; Hussein, T. Temporal Variations of Submicron Particle Number Concentrations at an Urban Background Site in Amman—Jordan. *Jordan J. Earth Environ. Sci.* **2019**, *10*, 37–44.
41. Abi-Esber, L.; El-Fadel, M. Indoor to outdoor air quality associations with self pollution implications inside passenger car cabins. *Atoms. Environ.* **2013**, *81*, 450–463. [CrossRef]
42. Alam, K.; Trautmann, T.; Blaschke, T.; Subhan, F. Changes in aerosol optical properties due to dust storms in the Middle East and Southwest Asia. *Remote Sens. Environ.* **2014**, *143*, 216–227. [CrossRef]
43. Roumie, M.; Chiari, M.; Srour, A.; Sa'adeh, H.; Reslan, A.; Sultan, M.; Ahmad, M.; Calzolai, G.; Nava, S.; Zubaidi, T.; et al. Evaluation and mapping of $PM_{2.5}$ atmospheric aerosols in Arasia region using PIXE and gravimetric measurements. *Nucl. Inst. Meth. Phys. Res. B* **2016**, *371*, 381–386. [CrossRef]
44. Hussein, T.; Alghamdi, M.A.; Khoder, M.; AbdelMaksoud, A.S.; Al-Jeelani, H.; Goknil, M.K.; Shabbaj, I.I.; Almehmadi, F.M.; Hyvärinen, A.; Lihavainen, H.; et al. Particulate matter and number concentrations of particles larger than 0.25 μm in the urban atmosphere of Jeddah, Saudi Arabia. *Aerosol Air Qual. Res.* **2014**, *14*, 1383–1391. [CrossRef]
45. Alghamdi, M.A.; Khoder, M.; Abdelmaksoud, A.S.; Harrison, R.M.; Hussein, T.; Lihavainen, H.; Al-Jeelani, H.; Goknil, M.H.; Shabbaj, I.I.; Almehmadi, F.M.; et al. Seasonal and diurnal variations of BTEX and their potential for ozone formation in the urban background atmosphere of the coastal city Jeddah, Saudi Arabia. *Air Qual. Atmos. Health* **2014**, *7*, 467–480. [CrossRef]
46. Alghamdi, M.A.; Khoder, M.; Harrison, R.M.; Hyvärinen, A.-P.; Hussein, T.; Al-Jeelani, H.; Abdelmaksoud, A.S.; Goknil, M.H.; Shabbaj, I.I.; Almehmadi, F.M.; et al. Temporal Variations of O_3 and NO_x in the Urban Background Atmosphere of the Coastal City Jeddah, Saudi Arabia. *Atmos. Environ.* **2014**, *94*, 205–214. [CrossRef]
47. Boman, J.; Shaltout, A.A.; Abozied, A.M.; Hassan, S.K. On the elemental composition of $PM_{2.5}$ in central Cairo, Egypt. *X-ray Spectrom.* **2013**, *42*, 276–283. [CrossRef]
48. Basha, G.; Phanikumar, D.V.; Kumar, K.N.; Ouarda, T.B.M.J.; Marpu, P.R. Investigation of aerosol optical, physical, and radiative characteristics of a severe dust storm observed over UAE. *Remote Sen. Environ.* **2015**, *169*, 404. [CrossRef]
49. Engelbrecht, J.P.; Jayanty, R.K.M. Assessing sources of airborne mineral dust and other aerosols, in Iraq. *Aeol. Res.* **2013**, *9*, 153–160. [CrossRef]
50. Engelbrecht, J.P.; McDonald, E.V.; Gillies, J.A.; Jayanty, R.K.M.; Casuccio, G.; Gertler, A.W. Characterizing Mineral Dusts and Other Aerosols from the Middle East—Part 1: Ambient Sampling. *Inhal. Toxicol.* **2009**, *21*, 297–326. [CrossRef]
51. Engelbrecht, J.P.; McDonald, E.V.; Gillies, J.A.; Jayanty, R.K.M.; Casuccio, G.; Gertler, A.W. Characterizing Mineral Dusts and Other Aerosols from the Middle East—Part 2: Grab Samples and Re-Suspensions. *Inhal. Toxicol.* **2009**, *21*, 327–336. [CrossRef]
52. Habeebullah, T.M. An Analysis of Air Pollution in Makkah—A View Point of Source Identification. *Environ. Asia* **2013**, *2*, 11–17.

53. Moustafa, M.; Mohamed, A.; Ahmed, A.-R.; Nazmy, H. Mass size distributions of elemental aerosols in industrial area. *J. Adv. Res.* **2015**, *6*, 827–832. [CrossRef] [PubMed]
54. Notaro, M.; Yu, Y.; Kalashnikova, O.V. Regime shift in Arabian dust activity, triggered by persistent Fertile Crescent drought. *J. Geophys. Res. Atmos.* **2015**, *120*, 10229–10249. [CrossRef]
55. Waked, A.; Afif, C.; Brioude, J.; Formenti, P.; Chevaillier, S.; El Haddad, I.; Doussin, J.-F.; Borbon, A.; Seigneur, C. Composition and Source Apportionment of Organic Aerosol in Beirut, Lebanon, During Winter 2012. *Aerosol Sci. Technol.* **2013**, *47*, 1258–1266. [CrossRef]
56. Waked, A.; Seigneur, C.; Couvidat, F.; Kim, Y.; Sartelet, K.; Afif, C.; Borbon, A.; Formenti, P.; Sauvage, S. Modeling air pollution in Lebanon: Evaluation at a suburban site in Beirut during summer. *Atmos. Chem. Phys.* **2013**, *13*, 5873–5886. [CrossRef]
57. Nazelle, A.; Fruin, S.; Westerdahl, D.; Mareinez, D.; Ripoll, A.; Kubesch, N.; Nieuwenhuijsen, M. A travel mode comparison of commuters' exposures to air pollutants in Barcelona. *Atoms. Environ.* **2012**, *59*, 151–159. [CrossRef]
58. Pérez, N.; Pey, J.; Cusack, M.; Reche, C.; Querol, X.; Alastuey, A.; Viana, M. Variability of Particle Number, Black Carbon, and PM_{10}, $PM_{2.5}$, and PM_1 Levels and Speciation: Influence of Road Traffic Emissions on Urban Air Quality. *Aerosol Sci. Technol.* **2010**, *44*, 487–499. [CrossRef]
59. Okokon, E.O.; Yli-Tuomi, T.; Turunen, A.W.; Taimisto, P.; Pennanen, A.; Vouitsis, I.; Samaras, Z.; Voogt, M.; Keuken, M.; Lanki, T. Particulates and noise exposure during bicycle, bus and car commuting: A study in three European cities. *Environ. Res.* **2017**, *154*, 181–189. [CrossRef]
60. Knibbs, L.D.; Cole-Hunter, T.; Morawska, L. A review of commuter exposure to ultrafine particles and its health effects. *Atoms. Environ.* **2011**, *45*, 2611–2622. [CrossRef]
61. Knibbs, L.D.; de Dear, R.J.; Morawska, L. Effect of cabin ventilation rate on ultrafine particle exposure inside automobiles. *Environ. Sci. Technol.* **2010**, *44*, 3546–3551. [CrossRef]
62. Lim, S.; Dirks, K.N.; Salmond, J.A.; Xie, S. Determinants of spikes in ultrafine particle concentration whilst commuting by bus. *Atoms. Environ.* **2015**, *112*, 1–8. [CrossRef]
63. Both, A.F.; Westerdahl, D.; Fruin, S.; Haryanto, B.; Marshall, J.D. Exposure to carbon monoxide, fine particle mass, and ultrafine particle number in Jakarta, Indonesia: Effect of commute mode. *Sci. Total Environ.* **2013**, *443*, 965–972. [CrossRef] [PubMed]
64. Zhang, Q.; Fischer, H.J.; Weiss, R.E.; Zhu, Y. Ultrafine particle concentrations in and around idling school buses. *Atoms. Environ.* **2013**, *69*, 65–75. [CrossRef]
65. Liu, L.J.S.; Phuleria, H.C.; Webber, W.; Davey, M.; Lawson, D.R.; Ireson, R.G.; Zielinska, B.; Ondov, J.M.; Weaver, C.S.; Lapin, C.A.; et al. Quantification of self pollution from two diesel school buses using three independent methods. *Atoms. Environ.* **2010**, *44*, 3422–3431.
66. Zhang, Q.; Zhu, Y. Measurements of ultrafine particles and other vehicular pollutants inside school buses in South Texas. *Atoms. Environ.* **2010**, *44*, 253–261. [CrossRef]
67. Dingenen, R.A.; Raes, F.; Putaud, J.-P.; Baltensperger, U.; Charron, A.; Facchini, M.-C.; Decesari, S.; Fuzzi, S.; Gehrige, R.; Hansson, H.-C.; et al. A European aerosol phenomenology—1: Physical characteristics of particulate matter at kerbside, urban, rural and background sites in Europe. *Atmos. Environ.* **2004**, *38*, 2561–2577. [CrossRef]
68. Salma, I.; Borsos, T.; Weidinger, T.; Alato, P.; Hussein, T.; Dal Maso, M.; Kulmala, M. Production, growth and properties of ultrafine atmospheric aerosol particles in an urban environment. *Atmos. Chem. Phys.* **2011**, *11*, 1339–1353. [CrossRef]
69. Ruuskanen, J.; Tuch, T.; Brink, H.T.; Peters, A.; Khlystov, A.; Mirme, A.; Kos, G.P.A.; Brunekreef, B.; Wichmann, H.E.; Buzorius, G.; et al. Concentrations of ultrafine, fine and $PM_{2.5}$ particles in three European cities. *Atmos. Environ.* **2001**, *35*, 3729–3738. [CrossRef]
70. Reche, C.; Querol, X.; Alastuey, A.; Viana, M.; Pey, J.; Moreno, T.; Rodriguez, S.; Gonzalez, Y.; Fernandez-Camacho, R.; de la Campa, A.M.S.; et al. New considerations for PM, Black Carbon and particle number concentration for air quality monitoring across different European cities. *Atmos. Chem. Phys.* **2011**, *11*, 6207–6227. [CrossRef]
71. Kumar, P.; Rivas, I.; Sachdeva, L. Exposure of in-pram babies to airborne particles during morning drop-in and afternoon pick-up of school children. *Environ. Pollut.* **2017**, *224*, 407–420. [CrossRef]

72. Ragettli, M.S.; Corradi, E.; Braun-Fahrländer, C.; Schindler, C.; de Nazelle, A.; Jerrett, M.; Ducret-Stich, R.E.; Künzli, N.; Phuleria, H.C. Commuter exposure to ultrafine particles in different urban locations, transportation modes and routes. *Atoms. Environ.* **2013**, *77*, 376–384. [CrossRef]
73. Panis, I.L.; de Geus, B.; Vandenbulcke, G.; Willems, H.; Degraeuwe, B.; Bleux, N.; Mishra, V.; Thomas, I.; Meeusen, R. Exposure to particulate matter in traffic: A comparison of cyclists and car passengers. *Atoms. Environ.* **2010**, *44*, 2263–2270. [CrossRef]
74. Quiros, D.C.; Lee, E.S.; Wang, R.; Zhu, Y. Ultrafine particle exposures while walking, cycling, and driving along an urban residential roadway. *Atoms. Environ.* **2013**, *73*, 185–194. [CrossRef]
75. Ham, W.; Vijayan, A.; Schulte, N.; Herner, J.D. Commuter exposure to $PM_{2.5}$, BC, and UFP in six common transport microenvironments in Sacramento, California. *Atoms. Environ.* **2017**, *167*, 335–345. [CrossRef]
76. Boarnet, M.G.; Houston, D.; Edwards, R.; Princevac, M.; Ferguson, G.; Pan, H.; Bartolome, C. Fine particulate concentrations on sidewalks in five Southern California cities. *Atoms. Environ.* **2011**, *45*, 4025–4033. [CrossRef]
77. Hankey, S.; Marshall, J.D. On-bicycle exposure to particulate air pollution: Particle number, black carbon, $PM_{2.5}$, and particle size. *Atoms. Environ.* **2015**, *122*, 65–73. [CrossRef]
78. Pattinson, W.; Longley, I.; Kingham, S. Using mobile monitoring to visualise diurnal variation of traffic pollutants across two near-highway neighbourhoods. *Atoms. Environ.* **2014**, *94*, 782–792. [CrossRef]
79. Goel, R.; Gani, S.; Guttikunda, S.K.; Wilson, D.; Tiwari, G. On-road $PM_{2.5}$ pollution exposure in multiple transport microenvironments in Delhi. *Atoms. Environ.* **2015**, *123*, 129–138. [CrossRef]
80. Apte, J.S.; Kirchstetter, T.W.; Reich, A.H.; Deshpande, S.J.; Kaushik, G.; Chel, A.; Marshall, J.D.; Nazaroff, W.W. Concentrations of fine, ultrafine, and black carbon particles in auto-rickshaws in New Delhi, India. *Atoms. Environ.* **2011**, *45*, 4470–4480. [CrossRef]
81. Qiu, Z.; Xu, X.; Song, J.; Luo, Y.; Zhao, R.; Zhou, B.X.W.; Hao, X.L.I.Y. Pedestrian exposure to traffic PM on different types of urban roads: A case study of Xi'an, China. *Sustain. Cities Soc.* **2017**, *32*, 475–485. [CrossRef]
82. Huang, J.; Deng, F.; Wu, S.; Guo, X. Comparisons of personal exposure to $PM_{2.5}$ and CO by different commuting modes in Beijing, China. *Sci. Total Environ.* **2012**, *425*, 52–59. [CrossRef] [PubMed]
83. Betancourt, R.M.; Galvis, B.; Balachandran, S.; Ramos-Bonilla, J.P.; Sarmiento, O.L.; Gallo-Murcia, S.M.; Contreras, Y. Exposure to fine particulate, black carbon, and particle number concentration in transportation microenvironments. *Atoms. Environ.* **2017**, *157*, 135–145. [CrossRef]
84. Kulmala, M.; Kontkanen, J.; Junninen, H.; Lehtipalo, K.; Manninen, H.E.; Nieminen, T.; Petaja, T.; Sipila, M.; Schobesberger, S.; Rantala, P.; et al. Direct Observations of Atmospheric Aerosol Nucleation. *Science* **2013**, *339*, 943–946. [CrossRef] [PubMed]
85. Mølgaard, B.; Hussein, T.; Corander, J.; Hämeri, K. Forecasting Size-Fractionated Particle Number Concentrations in the Urban Atmosphere. *Atmos. Environ.* **2012**, *46*, 155–163. [CrossRef]
86. Mølgaard, B.; Birmili, W.; Clifford, S.; Massling, A.; Eleftheriadis, K.; Norman, M.; Vratolis, S.; Wehner, B.; Corander, J.; Hämeri, K.; et al. Evaluation of a statistical forecast model for size-fractionated urban particle number concentrations using data from five European cities. *J. Aerosol Sci.* **2013**, *66*, 96–110. [CrossRef]
87. Hussein, T.; Karppinen, A.; Kukkonen, J.; Härkonen, J.; Aalto, P.P.; Hämeri, K.; Kerminen, V.M.; Kulmala, M. Meteorological dependence of size fractionated number concentrations of urban aerosol particles. *Atmos. Environ.* **2006**, *40*, 1427–1440. [CrossRef]
88. Hussein, T.; Kukkonen, J.; Korhonen, H.; Pohjola, M.; Pirjola, L.; Wriath, D.; Härkönen, J.; Teinilä, K.; Koponen, I.K.; Karppinen, A.; et al. Evaluation and modeling of the size fractionated aerosol particle number concentration measurements nearby a major road in Helsinki—Part II: Aerosol measurements within the SAPPHIRE project. *Atmos. Chem. Phys.* **2007**, *7*, 4081–4094. [CrossRef]
89. Harrison, R.; Yin, J. Particulate matter in the atmosphere: Which particle properties are important for its effects on health? *Sci. Total Environ.* **2000**, *249*, 85–101. [CrossRef]
90. Harrison, R.M.; Jones, A.M.; Barrowcliffe, R. Field study of the influence of meteorological factors and traffic volumes upon suspended particle mass at urban roadside sites of differing geometries. *Atmos. Environ.* **2004**, *38*, 6361–6369. [CrossRef]
91. Charron, A.; Harrison, R.M. Primary particle formation from vehicle emissions during exhaust dilution in the roadside atmosphere. *Atmos. Environ.* **2003**, *37*, 4109–4119. [CrossRef]
92. Gidhagen, L.; Johansson, C.; Langner, J.; Olivares, G. Simulation of NOx and ultrafine particles in a street canyon in Stockholm, Sweden. *Atmos. Environ.* **2004**, *38*, 2029–2044. [CrossRef]

93. Hosiokangas, J.; Vallius, M.; Ruuskanen, J.; Mirme, A.; Pekkanen, J. Resuspended dust episodes as an urban air-quality problem in subarctic regions. *Scand. J. Work Environ. Health* **2004**, *30* (Suppl. 2), S28–S35.
94. Järvi, L.; Hannuniemi, H.; Hussein, T.; Junninen, H.; Aalto, P.P.; Hillamo, R.; Mäkelä, T.; Keronen, P.; Siivola, E.; Vesala, T.; et al. The urban measurement station SMEAR III: Continuous monitoring of air pollution and surface-atmosphere interactions in Helsinki, Finland. *Boreal Environ. Res.* **2009**, *14* (Suppl. A), 86–109.
95. Krecl, P.; Ström, J.; Johansson, C. Diurnal variation of atmospheric aerosol during the wood combustion season in Northern Sweden. *Atmos. Environ.* **2008**, *42*, 4113–4125. [CrossRef]
96. Singh, A.K.; Rai, J.; Niwas, S. Variations of aerosols in relation to some meteorological parameters during different weather conditions. *Atmósfera* **2000**, *13*, 177–184.

Article

Characteristics and Source Apportionment of Metallic Elements in $PM_{2.5}$ at Urban and Suburban Sites in Beijing: Implication of Emission Reduction

Miaoling Li [1,2], Zirui Liu [1,*], Jing Chen [3], Xiaojuan Huang [1,4], Jingyun Liu [1,2], Yuzhu Xie [1], Bo Hu [1], Zhongjun Xu [2,*], Yuanxun Zhang [5] and Yuesi Wang [1,6]

1 State Key Laboratory of Atmospheric Boundary Layer Physics and Atmospheric Chemistry (LAPC), Institute of Atmospheric Physics, Chinese Academy of Sciences, Beijing 100029, China; limiaoling210@outlook.com (M.L.); xjhuang@cuit.edu.cn (X.H.); liujingyun@dq.cern.ac.cn (J.L.); xieyuzhu@dq.cern.ac.cn (Y.X.); hb@dq.cern.ac.cn (B.H.); wys@mail.iap.ac.cn (Y.W.)
2 Department of Environmental Science and Engineering, Beijing University of Chemical Technology, Beijing 100029, China
3 Beijing Meteorological Information Center, Beijing 100089, China; chenjinn@hotmail.com
4 Plateau Atmosphere and Environment Key Laboratory of Sichuan Province, School of Atmospheric Sciences, Chengdu University of Information Technology, Chengdu 610225, China
5 College of Resources and Environment, University of Chinese Academy of Sciences, Beijing 100049, China; yxzhang@ucas.edu.cn
6 Center for Excellence in Regional Atmospheric Environment, Institute of Urban Environment, Chinese Academy of Sciences, Xiamen 361021, China
* Correspondence: liuzirui@mail.iap.ac.cn (Z.L.); xuzj@mail.buct.edu.cn (Z.X.)

Received: 24 January 2019; Accepted: 20 February 2019; Published: 26 February 2019

Abstract: To gain insights into the impacts of emission reduction measures on the characteristics and sources of trace elements during the 2014 Asia-Pacific Economic Cooperation (APEC) summit, $PM_{2.5}$ samples were simultaneously collected from an urban site and a suburban site in Beijing from September 15th to November 12th, and fifteen metallic elements were analyzed, including five crustal elements (Mg, Al, K, Ca and Fe), nine trace metals (V, Cr, Mn, Co, Cu, Zn, Ag, Cd and Pb) and As. Most of the trace metals (V, Cr, Mn, As, Cd and Pb) decreased more than 40% due to the emission regulations during APEC, while the crustal elements decreased considerably (4–45%). Relative to the daytime, trace metals increased during the nighttime at both sites before the APEC summit, but no significant difference was observed during the APEC summit, suggesting suppressed emissions from anthropogenic activities. Five sources (dust, traffic exhaust, industrial sources, coal and oil combustion and biomass burning) were resolved using positive matrix factorization (PMF), which were collectively decreased by 30.7% at the urban site and 14.4% at the suburban site during the APEC summit. Coal and oil combustion regulations were the most effective for reducing the trace elements concentrations (urban site: 63.1%; suburban site: 52.0%), followed by measures to reduce traffic exhaust (52.8%) at the urban site and measures to reduce biomass burning (37.7%) at the suburban site. Our results signify that future control efforts of metallic elements in megacities like Beijing should prioritize coal and oil combustion, as well as traffic emissions.

Keywords: Asia-Pacific Economic Cooperation summit; $PM_{2.5}$; trace elements; source apportionment

1. Introduction

$PM_{2.5}$, also known as atmospheric fine particles, refers to particulate matter with an aerodynamic diameter of less than or equal to 2.5 μm. These particles not only endanger human health but also impact regional atmospheric visibility and environmental climate [1–3]. Because $PM_{2.5}$ particles are small in size and have large specific surface areas, they easily adsorb large quantities of organic compounds, such as mutagens, pathogens and heavy metals. Trace elements are some of the most important components in $PM_{2.5}$, as they are non-biodegradable, highly enriched and highly toxic. Previous studies have found that Mn, Ni, Cu, Zn, Se and Pb are potentially toxic to the human body, and that As, Cr and Ni are potent carcinogens that can seriously threaten human health [4–7]. Besides the effects on human health, trace metals could also damage ecological environments through dry and wet deposition to terrestrial and aquatic ecosystems [8].

The Asia-Pacific Economic Cooperation (APEC) summit that convened in November 2014, which is the transition period between autumn and winter, was held in Beijing. During this period, the air was dry, the wind speed was low, and the static inversion temperature was relatively stable. Beijing is surrounded by mountains on three sides (to the north, south and west), and thus the southerly airflow near the ground is weak and is consequently not conducive for the horizontal diffusion of pollutants. In addition, the strong sea level pressure and uniform pressure field that collectively result in the vertical stability of the atmospheric stratification over Beijing are not conducive for the vertical diffusion of pollutants. Moreover, the Beijing-Tianjin-Hebei region contains numerous highly polluting industries, including the steel, cement, oil refining, and petrochemical industries, which emit enormous quantities of air pollutants that are carried along southerly currents from Hebei, Tianjin, and Shandong toward Beijing. To ensure a desirable level of air quality during the APEC conference, Beijing, Tianjin, Hebei, Shanxi, and Shandong Provinces in addition to Inner Mongolia and other provinces and cities have adopted emission reduction measures of unprecedented intensity and scope, thereby generating the characteristic "APEC blue" sky over Beijing during the summit. These control measures include regulations permitting only even or odd license plate number vehicle use, air pollution supervision for air quality, vacation days off for the public, construction site restrictions to mitigate road dust, polluting company and industry restrictions, and regional controls by several provinces.

Several studies were performed within the Beijing-Tianjin-Hebei region regarding the air quality and impacts of emission regulations on $PM_{2.5}$ during the APEC summit [9–12]. A previous study also examined the characteristics of $PM_{2.5}$, including the concentrations, size distributions and sources of $PM_{2.5}$, during the APEC summit in Beijing [13]. The relative impacts of emission regulations and meteorological properties on the mitigation of air pollution over Beijing during the APEC summit were also studied [14], as were the source apportionment of atmospheric ammonia before, during, and after the conference [15]. However, these previous studies have rarely considered the effects of control measures on the trace elements in fine particles, which was more associated with the human health. In addition, although the characteristics and sources of trace elements in $PM_{2.5}$ have been studied at length [16–18], previous studies regarding the properties of trace elements had not been conducted under different emission scenarios, especially for the emission reduction scenario in the near future. The APEC summit provided a unique city-wide experiment to examine the response of various trace element sources to such comprehensive and intensive mitigation efforts.

In the present study, the pollution characteristics of metallic elements in $PM_{2.5}$ before and after the 2014 APEC summit were analyzed by simultaneously measuring the mass concentrations of metallic elements in the urban and suburban areas of Beijing. An analysis of the enrichment factor (*EF*) and a positive matrix factorization (PMF) model were used to compare the sources of metal elements due to emissions reduction. The results of these analyses could be used to understand the effects of emission reduction measures on pollution levels and the pollution characteristics of metal elements and to provide suggestions for future air pollution regulations.

2. Materials and Methods

2.1. Sites and Sampling

The $PM_{2.5}$ sampling was conducted simultaneously at the urban and suburban sites in Beijing, with two $PM_{2.5}$ samplers (TH-150C, Tianhong, Wuhan) at an airflow rate of 100 L min^{-1} from 15 September to 12 November 2014. The urban site (JDM) was established in the Institute of Atmospheric Physics (IAP), Chinese Academy of Sciences (39.97° N, 116.38° E), which is located between the North Third Ring Road and Fourth Ring Road in Beijing. This site is approximately 1 km from the Third Ring Road, 200 m away from the north-south highway, and 50 m north of East-West Beitucheng West Road. The sampler was situated on the rooftop of a two-floor building, approximately 10 m above the ground. The suburban site (HR) was established on the campus of the University of Chinese Academy of Sciences (40.41° N, 116.68° E), approximately 55 km northeast of the urban site (JDM). This site is located at the foot of the Great Wall adjacent to Yanqi Lake. The sampler was placed on rooftop of a four- floor building approximately 15 m above the ground. Two $PM_{2.5}$ samples were collected each day at both sites, one from 08:00 to 19:30 (daytime measurement) and one from 20:00 to 07:30 the next day (nighttime measurement). In total, 116 samples were acquired at each site. After sampling, all the filters were sealed and stored in a refrigerator (−20 °C) until subsequent analysis. An automatic meteorological observation instrument (Vaisala Company, Finland) was used to observe various meteorological factors at the same site, including the sea level pressure, ambient temperature, relative humidity, wind speed and wind direction.

2.2. Chemical Analysis

The quartz fiber filters packaged with aluminum foil were pre-fired in a Muffle furnace at 500 °C for 4 h to remove a small amount of organic matter. To reduce the influences of volatilization and water vapor during the weighing process, the filters were balanced within a chamber at constant temperature and constant humidity (temperature: 20 °C ± 1 °C; humidity: 40% ± 5%) for more than 48 h both before and after sampling. Then, these filters were weighed using a microelectronic balance with a reading precision of 10 μg. To guarantee the accuracy of the weights, the weighing was repeated until a difference of less than 0.10 mg between the two measured weights was achieved. The net increase in the quality of the filter from before sampling to after sampling was used to represent the quality of the particulate matter. The mass of the particles was divided by the volume of the sample to obtain the atmospheric $PM_{2.5}$ masses.

In this study, microwave acid digestion was used to digest the filter samples into a liquid solution for elemental analysis. One-quarter of each filter sample was placed in the digestion vessel with a mixture of 6mL HNO_3, 2mL H_2O_2, and 0.6mL HF and was then exposed to a three-stage microwave digestion procedure from a microwave-accelerated reaction system (MARS; CEM Corporation, USA). After that, the digestion solution was transferred to polyethylene terephthalate (PET) bottles and diluted to 50 mL with deionized water (with a conductivity: 18.2 MΩ/cm). Inductively coupled plasma mass spectrometry (ICP-MS 7500a; Agilent Technologies, Japan subsidiary) was used to determine the concentrations of 15 trace elements in the digestion solution, including Mg, Al, K, Ca, V, Cr, Mn, Fe, Co, Cu, Zn, As, Ag, Cd and Pb. The concentrations of the trace elements were quantified using a multi-element external standard, and the external standard liquid was diluted into four concentration gradients with 5% HNO_3. The correlation coefficient of the standard curve exceeded 0.9999. The internal standard elements solutions (^{45}Sc, ^{72}Ge, ^{103}Rh, ^{115}In, ^{159}Tb, ^{175}Lu and ^{209}Bi) were simultaneously placed into the instrument with the samples during the analyses. Each sample was determined 3 times. The relative standard deviation (RSD) of the internal standard element was less than 3%, indicating that the instrument was stable. When the RSD was greater than 3%, the sample was analyzed again. The concentrations of the elements in $PM_{2.5}$ were calculated by measuring the concentrations of the elements and the sampling volume. More detailed information, such as instrument optimization, calibration, and quality control, is given in [19].

2.3. Data Analysis

2.3.1. Enrichment Factor (EF)

The *EF* is often used to measure the enrichment of elements in atmospheric particulate matter and to determine and estimate the natural and anthropogenic sources of those metallic elements. The *EF* is calculated as follows:

$$EF = \frac{(C_\alpha / C_\beta)_{aerosol}}{(C_\alpha / C_\beta)_{crust}} \quad (1)$$

where C_α is the mass concentration of the investigated element α, C_β is the mass concentration of a comparable element β, $(C_\alpha / C_\beta)_{aerosol}$ is the ratio of comparable elements to reference elements in the aerosol, and $(C_\alpha / C_\beta)_{crust}$ is the ratio of research elements to comparable elements in the crust. Usually, the elements that are relatively more stable, more volatile, less anthropogenic and ubiquitous throughout the crust are selected as comparable elements (Al, Fe, Ti, Sc, and Si are commonly chosen). In this study, Al was chosen as the comparable element. In addition, the concentrations of the crustal elements were derived from a previous investigation [20]. When the *EF* is less than 10, the element is mainly derived from the crust, while *EF* values between 10 and 100 indicate that the element originated from both natural and anthropogenic sources. When the *EF* is greater than 100, the element mainly originated from anthropogenic sources [21].

2.3.2. Positive Matrix Factorization (PMF)

The PMF model, which is based on the factor analysis method, was first developed in 1993 by Paatero [22], after which it was quickly popularized and broadly applied. Compared with chemical mass balance (CMB) models, PMF models do not need to analyze information regarding the pollution source. In addition, it is not necessary to establish the pollution source composition spectrum; rather, the various factors can be obtained by using the standard deviation of the data [23]. The basic principle of this model is to assume that X is an $n \times m$ matrix, where n is the number of samples and m is the number of chemical components.

$$x_{ij} = \sum_{k=1}^{p} g_{ik} f_{kj} + e_{ij} \quad (i = 1,2\ldots n; j = 1,2\ldots m; k = 1,2\ldots p) \quad (2)$$

The goal of the optimization is that the objective function Q will tend toward the value of the degrees of freedom.

$$Q(E) = \sum_{i=1}^{n} \sum_{j=1}^{m} (e_{ij} / u_{ij})^2 \quad (3)$$

where x_{ij} is the concentration of a substance j in the receptor on day i, g_{ik} is the contribution of a factor k to the receptor on day i, f_{kj} is the fraction of the chemical composition of a substance j of the factor k, e_{ij} is the residual of the chemical composition of a substance j on day i, and u_{ij} is the uncertainty of X.

Missing data values were substituted with median concentrations. The uncertainty (*Unc*) is calculated according to the method recommended by the U.S. Environmental Protection Agency (EPA) PMF Fundamentals [24–27].

When the concentration of the chemical composition is lower than the method detection limit (*MDL*) of the instrument, the uncertainty is as follows:

$$Unc = \frac{5}{6} MDL \quad (4)$$

When the concentration of the chemical composition is higher than the *MDL* of the instrument, the uncertainty is as follows:

$$Unc = \sqrt{(ErrorFraction + Concentration)^2 + (MDL)^2} \quad (5)$$

In this study, we uses the multivariate EPA PMF 5.0 model to resolve the sources of metallic elements. The daily $PM_{2.5}$ elemental composition datasets obtained at the two sites during the entire observation period (15 September to 12 November) were applied to PMF 5.0 to determine sources and their contributions to the total metallic element mass. The identification of the sources was based on certain chemical tracers that are generally presumed to be emitted by specific sources and are present in significant amounts in the samples collected. After testing the PMF results between 4 and 9 factors, the five factors solution was chosen by comparing PMF factor profiles with reference source profiles and tracers from previous studies. For the selected 5-factor case, the G-space plot pairings showed that the points were distributed across the solution space between the axes. Based on the Displacement (DISP) and Bootstrap-Displacement (BS-DISP) results, the FPEAK value set at 0.1 (JDM) and −0.1 (HR).

3. Results and Discussion

3.1. $PM_{2.5}$ Mass Concentration and Meteorological Conditions

The meteorological conditions and $PM_{2.5}$ mass concentrations obtained from the observation period are given in Figure 1. The average $PM_{2.5}$ mass concentrations before the APEC summit (BAPEC period, from 15 September to 2 November) were 127.6 $\mu g/m^3$ (JDM) and 87.7 $\mu g/m^3$ (HR), both of which exceed the second level of the National Ambient Air Quality Standard (75 $\mu g/m^3$). The numbers of days with average mass concentrations that surpassed the standard level of 75 $\mu g/m^3$ were 30 and 20 days at JDM and HR, respectively, accounting for 63.8% and 42.6% of the total number of sampling days. If the mass concentration of $PM_{2.5}$ exceeds 75 for two consecutive days, it is classified as a pollution event in this study. During the BAPEC period, 7 (JDM) and 5 (HR) pollution events were observed. Prior to each pollution event, southerly wind flows with short durations were usually observed. In addition, stagnant weather is usually accompanied by faint southern and southeastern winds [28], which would benefit the transportation of atmospheric pollutants to Beijing from its adjacent southern areas where the intensive industrial zone were located. As showed in Figure 1, a 6-day pollution event was observed from 6 October to 11 October at the urban and suburban sites. At the urban site (JDM), a continuous southerly wind with a mean wind speed of 1.4 m/s was observed from 12:00 to 24:00 on October 6th. Then, the wind speed largely decreased, and the average wind speed decreased to 0.3 m/s in the following 4 days. In addition, the average relative humidity increased to 71.2%. These conditions were not conducive to the dispersal of pollutants; thus, the $PM_{2.5}$ concentration reached a maximum of 428 $\mu g/m^3$ on 9 October. On the afternoon of 11 October, a northerly wind with an average wind speed of 2.2 m/s was observed, following which the pollution gradually dissipated. During the same pollution event, the average $PM_{2.5}$ mass concentration at the suburban area (HR) was 195.8 $\mu g/m^3$ with a maximum value of 271.3 $\mu g/m^3$, much lower than that at the urban area (JDM). Apart from the lower anthropogenic emissions in the suburban site, the higher average wind speed (1.1 m/s) at HR suggest a better diffusion condition and leading to the lower $PM_{2.5}$.

As shown in Table 1, the average $PM_{2.5}$ mass concentrations at sites JDM and HR during the APEC summit were 48.4 $\mu g/m^3$ and 33.1 $\mu g/m^3$, respectively, both of which were significantly lower than those during the BAPEC period ($p < 0.05$). The numbers of days with pollution levels that exceeded 75 $\mu g/m^3$ were only 2 and 1 at JDM and HR, respectively, accounting for 16.7% and 8.3% of the total number of sampling days. In addition, no heavy pollution days were observed at either site. Compared with the BAPEC period, the wind speeds increased and the relative humidity decreased during the APEC summit in the urban site, which suggested the meteorological conditions were conducive to the dispersal of pollutants, and the favorable meteorological conditions also played an important role in maintaining good air quality during the APEC summit. However, it should be noted that the wind speeds was comparable during BAPEC and APEC periods in the suburban site, which may suggest a similar meteorological conditions and the control measures would be more important that contributed to the good air quality.

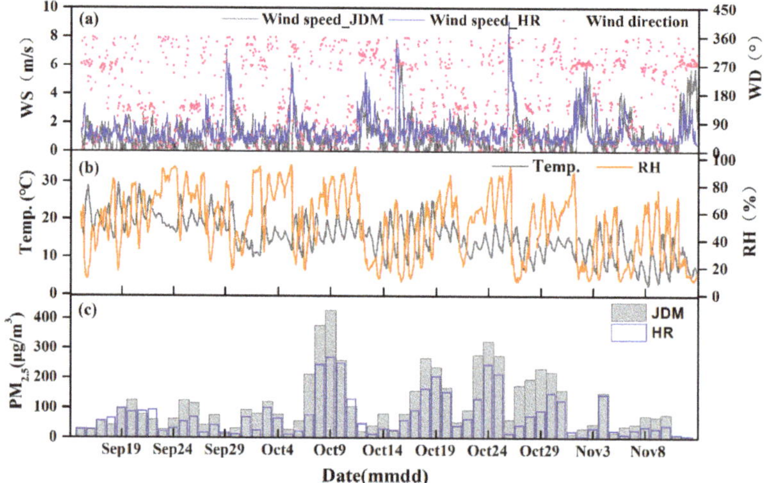

Figure 1. (**a**,**b**) Meteorological conditions and (**c**) $PM_{2.5}$ mass concentration during the sampling period at the Beijing urban site (JDM) and suburban site (HR).

Table 1. $PM_{2.5}$ mass concentrations and meteorological parameters before and during the Asia-Pacific Economic Cooperation (APEC) summit in Beijing urban site (JDM) and suburban site (HR).

Parameters	JDM (*n* = 47)		HR (*n* = 12)		Reduction (%)	
	BAPEC	APEC	BAPEC	APEC	JDM	HR
$PM_{2.5}$ (μg/m^3)	127.6	48.4	87.7	33.1	62.1 *	62.3 *
Days of $PM_{2.5}$ exceeding 75 μg/m^3	30	2	20	1	93.3	95.0
Exceeding Rate (%)	63.8	16.7	42.6	8.3	73.8	80.5
Temperature (°C)	16.8	10.1	14.3	6.9	39.9 **	21.7 **
Relative Humidity (%)	56.3	34.1	69.5	43.9	39.4 **	36.8 **
Wind Speed (m/s)	0.82	1.58	1.44	1.36	−92.7 **	5.6

* denote the difference reach significance level ($p < 0.05$), and ** denote the significance level ($p < 0.01$).

3.2. Concentrations of Metallic Elements

Table 2 presents the average mass concentrations of metallic elements in $PM_{2.5}$ in each site during the BAPEC and APEC periods. In both the urban and suburban sites, the concentrations of K, Fe, Ca, Al and Mg were the highest among the metallic elements examined herein. These 5 elements are all naturally occurring within the crust, and their total concentration accounted for more than 85% of the total concentration of all of the measured metallic elements. The concentration of Cd was less than 5 ng/m^3, which is the reference concentration limit provided by the standard GB3095-2012 and the World Health Organization (WHO). The concentration of As during the BAPEC period exceeded the reference limit (6 ng/m^3) provided by the GB3095-2012 and WHO standards, while that during APEC was less than 6 ng/m^3. The concentration of Pb was less than the seasonal concentration limit (1100 ng/m^3) specified by GB3095-2012 and was also less than the mean annual concentration limit (500 ng/m^3). The concentrations of crustal elements in Beijing decreased after 2010 relative to those prior to 2010; the concentrations of Mg, Al and Ca decreased by approximately 30% while those of K, Fe and Mn decreased by approximately 60% (Table S1). Likely as a result of the growth in car ownership throughout Beijing in recent years, the Cu concentration rose by 67.1%, especially since a large quantity of Cu is produced during car manufacturing and braking. The concentration of Pb decreased by 34.4% due to the increased usage of unleaded gasoline during recent years; interestingly, the Pb concentration decreased in spite of the aforementioned increase in car ownership.

The concentrations of V, As, Cd and Cr, which are related to industrial sources and coal-fired fuel sources, decreased by more than 70%. These decreases may have been related to the relocation of the Capital Steel Corporation Limited in 2010.

During the BAPEC period, the majority of the observed metallic elements were significantly higher ($p < 0.05$) in the urban site than in the suburban site, except for the Mg and Ag (Table 2). The same phenomenon was also observed during APEC, however, the difference in concentration of Al, Mn, Cd and Pb were insignificant. In addition, another two more element (As and Zn) showed higher concentrations at the suburban site during APEC. The concentration of each element was greater during the BAPEC period than during the APEC, and the metallic element concentrations were greater at JDM than at HR, regardless of the date. The total concentrations of metallic elements during the BAPEC period were 5028.1 ng/m^3 and 2851 ng/m^3 at JDM and HR, respectively, which were approximately 1.4 times the total site concentrations during the APEC. Compared with the BAPEC period, all the metallic elements in $PM_{2.5}$ decreased, and a significant reduction ($p < 0.05$) was observed for Mg, K, V, Co, Zn, As, Ag, Cd and Pb at the urban site. For the suburban site, although higher reductions of Ca, Cr, Fe, Co, Cu, Cd and Pb were observed compared with the urban site, the other elements showed insignificant reduction, especially for the Ag and Cu, which may suggested the different effects of control measures present in the urban and suburban areas in Beijing.

Table 2. Average mass concentrations of metallic elements in $PM_{2.5}$ and the differences between urban and suburban site and the reductions between before APEC (BAPEC) and APEC periods.

	BAPEC (n = 47)			APEC (n = 12)			Reduction (%) [b]	
ng/m^3	JDM	HR	Diff. (%) [a]	JDM	HR	Diff. (%) [a]	JDM	HR
Mg	160.7	167.6	−4.3	88.3	142.6	−61.5 *	45.1 **	14.9
Al	501	383.1	23.5 **c	416.2	365.2	12.3	16.9	4.7 *
K	1789.9	718.6	59.9 **	1125.3	460.3	59.1 **	37.1 *	35.9 *
Ca	641.9	456.3	28.9 **	616.5	369.9	40.0 **	4.0	18.9
V	2.6	1.2	53.8 **	0.76	0.36	52.6 **	70.8 **	70.0 **
Cr	9.5	4.0	57.9 **	6.5	2.6	60.0 **	31.6	35.0 **
Mn	59.3	37.4	36.9 **	31.3	20.6	34.2	47.2 **	44.9 **
Fe	1360.9	766.6	43.7 **	1139.1	455.4	60.0 **	16.3	40.6 **
Co	0.59	0.31	47.5 **	0.36	0.17	52.8 **	39.0 *	45.2
Cu	49.8	23.1	53.6 **	45.2	16.3	63.9 **	9.2	29.4
Zn	291.8	180.7	38.1 **	130.1	146.8	−12.8	55.4 **	18.8
As	10.0	9.5	5.0	4.0	4.7	−17.5	60.0 **	50.5 *
Ag	0.53	0.67	−26.4 *	0.35	0.67	−91.4 **	40.0 *	0.0
Cd	2.6	2.0	23.1 *	1.2	0.9	25.0	53.8 **	55.0 *
Pb	147.1	99.9	32.1 **	62.5	41.1	34.2	57.5 **	58.9 **

[a] The difference (Diff.) in metallic elements concentrations between urban (JDM) and suburban (HR) site; [b] The reduction in metallic elements concentrations between the BAPEC and APEC periods; [c],* denote the difference reach significance level ($p < 0.05$), and ** denote the significance level ($p < 0.01$).

3.3. Day/Night Variations in Metallic Elements

Figures 2 and 3 display the daytime and nighttime concentrations and *EF*s, respectively, of metallic elements during the BAPEC period and during the APEC. Compared with that during the daytime, the total concentration of metallic elements at night increased at JDM and declined at HR, regardless of the date. During the BAPEC period at JDM, the concentration of each crustal element (except K) decreased at night; the concentration and *EF* of K respectively increased by 24.7% and from 12.1 in the daytime to 13.7 at night, which suggests that K could be influenced by human activities like biomass burning that usually conducted in the south area of Beijing after the autumn harvest. The concentrations and *EF*s of the crustal elements at HR were generally reduced at night; their *EF*s were all less than 10, suggesting that the crustal elements at HR were not affected by anthropogenic factors. During the APEC, the variations in the crustal element concentrations at the two sites were similar to those during the BAPEC period.

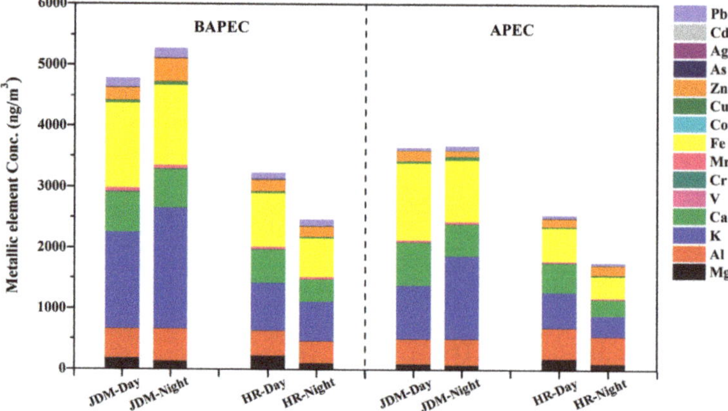

Figure 2. Daytime and nighttime concentrations of metallic elements before and during the APEC summit in Beijing urban site (JDM) and suburban site (HR).

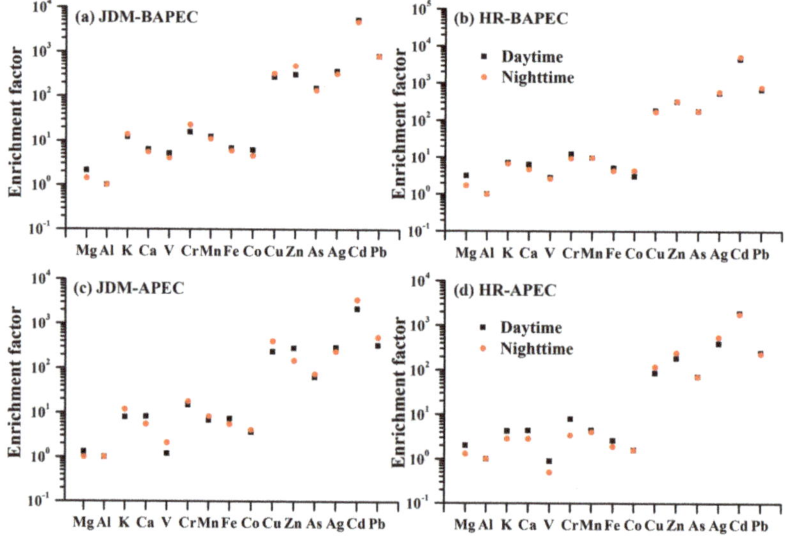

Figure 3. Daytime and nighttime enrichment factors of metallic elements (**a**,**b**) before and (**c**,**d**) during the APEC summit in Beijing urban site (JDM) and suburban site (HR).

During the BAPEC period, the total concentration of trace metal elements increased from 422.0 ng/m^3 during the daytime to 608.9 ng/m^3, during the nighttime at JDM, representing an increase of 44.3%. The concentrations of Cr, Cu, Zn and Pb increased by 58.6%, 32%, 72.4% and 10%, respectively. Moreover, the *EF*s of Cu and Zn rose from 266.3 and 306.4 (daytime) to 320.1 and 481.0 (nighttime), respectively. The *EF*s of Cu and Zn increased significantly at night. The daytime and nighttime concentrations of metallic elements within the $PM_{2.5}$ pollution at JDM during the BAPEC period were significantly different. This shows that the Cr, Cu, Zn and Pb detected at JDM were heavily affected by anthropogenic factors during the nighttime. A previous study found that the nighttime concentrations of organic carbon (OC) and elemental carbon (EC) were higher than those during the daytime, which was primarily due to lower temperatures and increased carbon emissions at night [29]. In this study, the concentrations of Cu, Zn and Pb increased significantly at night, reflecting enhanced levels of traffic activity. Moreover, the atmospheric boundary layer was generally lower and the wind speed was reduced (from 1.0 m/s to 0.6 m/s) at night, which could have worsened diffusion conditions and increased the concentrations of metallic elements. During the APEC summit, the total concentration of trace metal elements at JDM at night was 242 ng/m^3, which was 6.9% lower than that during the daytime; the variation in the concentration and *EF*s of each element was not obvious, however, as their differences were minor. This suggests that the regional emission control measures played a dominant role in driving these diurnal variations in urban area in Beijing.

In contrast, the concentrations of the trace metal elements at night decreased during the BAPEC period at HR. The concentrations of V, Cr, Cu, Zn and As decreased by 10.3–34.8% during the nighttime, indicating that biomass burning, motor vehicle activity and coal burning decreased during the nighttime at HR. In addition, increased wind speeds at night would have accelerated the diffusion of pollution, thereby reducing the impact of human factors on the pollution at HR at night. Interestingly, the concentrations of Cu, Zn and Ag increased at night by 20.5%, 16% and 21.5%, respectively, during the APEC at HR, and the *EF*s of Cu and Zn also increased slightly. This indicates that the effects of anthropogenic factors on Cu and Zn were enhanced at night during the APEC summit in contrast to the diurnal trends during the BAPEC period. Therefore, the control measures during daytime are more effective than those at the night in the suburban area of Beijing during the APEC summit.

3.4. Characteristics of Metallic Elements under Different Pollution Levels

The air quality was classified into three categories based on the second level of the National Ambient Air Quality Standard (GB3095-2012): when the concentration was less than 75 μg/m^3, the air quality was classified as clean days (C); when the mass concentration was between 75 and 150 μg/m^3, the air quality was classified as light polluted days (L); and when the concentration exceeded 150 μg/m^3, the air quality was classified as heavy polluted days (H).

The concentrations and *EF*s of the metallic elements in different pollution levels both during the BAPEC period and during the APEC summit are shown in Figures 4 and 5, respectively. Generally, the total concentration of metallic elements increased with an increase in the pollution at either site. The proportion of each metallic element in $PM_{2.5}$ decreased, indicating that the main source of the increase in $PM_{2.5}$ was not the growth of metallic elements, which is consistent with the findings of a previous study [30]. The concentration of the majority metallic elements was increased with relatively aggravated pollution levels during the BAPEC in the urban site, and the significant increase ($p < 0.05$) was observed for K, Cr, Mn, Fe, Co, Cu, Zn, As, Ag, Cd and Pb when pollution level changed from clean days to heavily polluted days and there no significant difference for all the 15 elements when pollution level changed from clean days to lightly polluted days. Interestingly, the concentrations of Mg, Al and Ca varied insignificantly and irregularly during the BAPEC period, coincident with relatively aggravated pollution levels. Similar variation of the metallic elements under different pollution levels was also observed in the suburban site. The majority of trace metal elements were significantly increased when the pollution level changed from clean days to heavily polluted days, while the crustal elements like Mg, Al and Ca showed an inconspicuous increase. Meanwhile, the *EF*s of the majority metallic elements was increased with relatively aggravated pollution levels, except for the Co and Ag in the suburban site. For example, the *EF* of K was less than 10 on clear days and higher than 10 on pollution days, indicating that K is more affected by human factors on pollution days. The concentration of As on severely polluted days was 3.6 times higher than that on clear days and 1.9 times higher than that on light pollution days. The concentrations, *EF*s and proportions of V, Co, Ag and Pb to the total metallic elements all exhibited their maximum values on light pollution days at HR. The concentrations of other trace metal elements, such as As and Cd, increased sharply with an increase in the pollution severity; their concentrations were 3.5 times higher on heavy pollution days than on clear days.

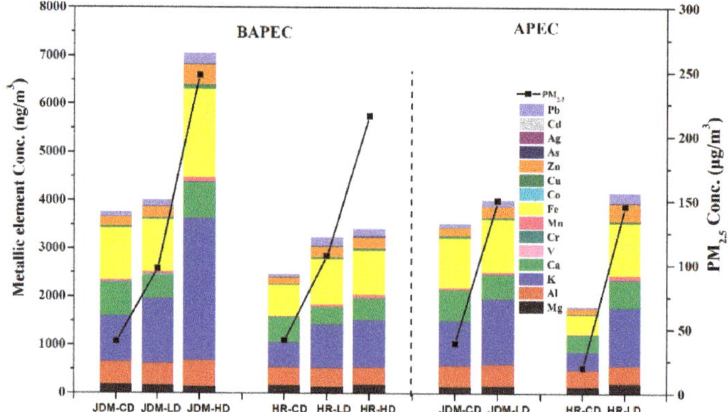

Figure 4. Concentrations of metallic elements at clean days (CD), light polluted days (LD) and heavy polluted days (HD) before and during the APEC summit in Beijing urban site (JDM) and suburban site (HR).

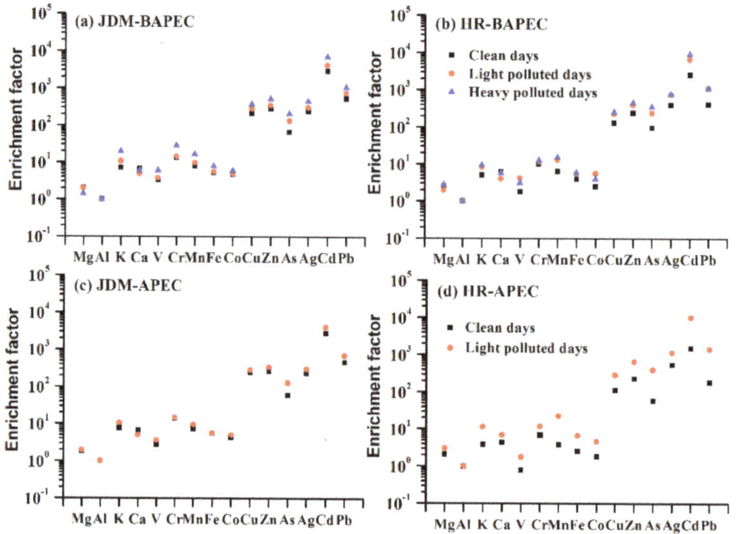

Figure 5. Enrichment factors of metallic elements on clean days, light polluted days and heavy polluted days (**a**,**b**) before and (**c**,**d**) during the APEC summit in Beijing urban site (JDM) and suburban site (HR).

During APEC, there were only two and one lightly polluted days at the urban and suburban sites respect tively, and no heavily polluted day at either site was observed. The total metallic element concentration on pollution days was 2.3 times higher than that on clear days at HR and 1.1 times higher at JDM. With the exception of Ca, the crustal elements at JDM exhibited magnitudes of increase; for example, the concentration of K rose from 925.4 ng/m^3 to 1344.9 ng/m^3. The concentrations of the crustal elements Mg, Al and Ca on lightly polluted days were 1~1.7 times those on clear days at HR. The elements K, Mn and Fe showed broader ranges of increases ranging from 2.8 to 6.3 times greater concentrations on pollution days. Among them, the *EF* of Mn increased from 3.8 to 22.7, indicating that anthropogenic emissions of Mn was largely enhanced on pollution days. With an increase in the pollution, the concentrations and *EF*s of trace metal elements at the two sites increased, and the range in the increased concentrations at HR was broader than that at JDM. The concentrations of As, Cd and Pb on lightly polluted days were 1.5~2.3 and 6.9~8 times higher than those on clear days at JDM and HR, respectively. Furthermore, the *EF*s of Mn, Fe, Cu, Zn, As, Cd and Pb on polluted days were 1.3~2.2 and 2.7~7.8 times higher than those on clear days at JDM and HR, respectively. These results indicates that traffic exhaust, industrial sources and coal combustion may be the main sources of trace metal elements, which is consistent with the findings of previous research [31]. Therefore, trace metal elements are more easily enriched on pollution days.

3.5. Source Apportionment of Metallic Elements

The source profiles and the time series of the factors are displayed in Figures 6 and 7. The percentages of each source are shown in Figure 8. The PMF source factors for the metallic elements for the entire study and during the BAPEC and APEC periods, in addition to the changes in their percentages during the APEC summit are provided in Table 3. The five sources of metallic elements, that is, dust, traffic exhaust, industrial sources, coal and oil combustion and biomass burning, were identified. Secondary aerosols is not identified by this study, which was different from the previous source appointment studies conducted in Beijing (Table S2). It should be note that previous source appointment studies were mainly concerned with the total $PM_{2.5}$ mass, while this study was concerned about the metallic elements, which may lead to the different source contribution results between this study and the prior ones.

Table 3. Averaged concentration of metallic elements in $PM_{2.5}$ for the five factors for before and during the APEC summit in Beijing urban site (JDM) and suburban site (HR).

Sources	JDM			HR		
	BAPEC	**APEC**	**Reduction (%)**	**BAPEC**	**APEC**	**Reduction (%)**
Dust (ng/m^3)	162.0	148.1	−8.6	95.5	122.7	28.5
Traffic Exhaust (ng/m^3)	100.2	47.3	−52.8	65.8	43.6	−33.7
Industrial Emissions (ng/m^3)	83.0	59.9	−27.9	46.8	31.6	−32.4
Coal and Oil Combustion (ng/m^3)	43.7	16.1	−63.1	23.5	11.3	−52.0
Biomass Burning (ng/m^3)	101.0	68.4	−32.2	47.3	29.5	−37.7

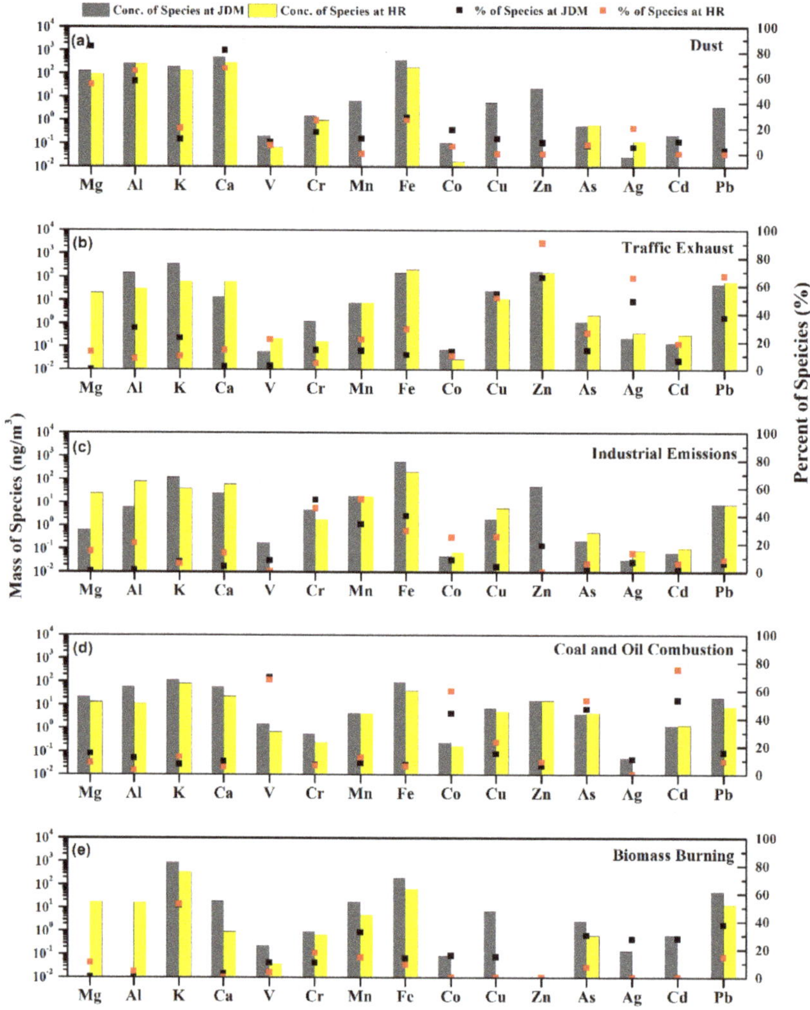

Figure 6. Source profile of the five factors ((**a**) dust, (**b**) traffic exhaust, (**c**) industrial sources, (**d**) coal and oil combustion and (**e**) biomass burning) of metallic elements in $PM_{2.5}$ resolved by PMF in Beijing urban site (JDM) and suburban site (HR).

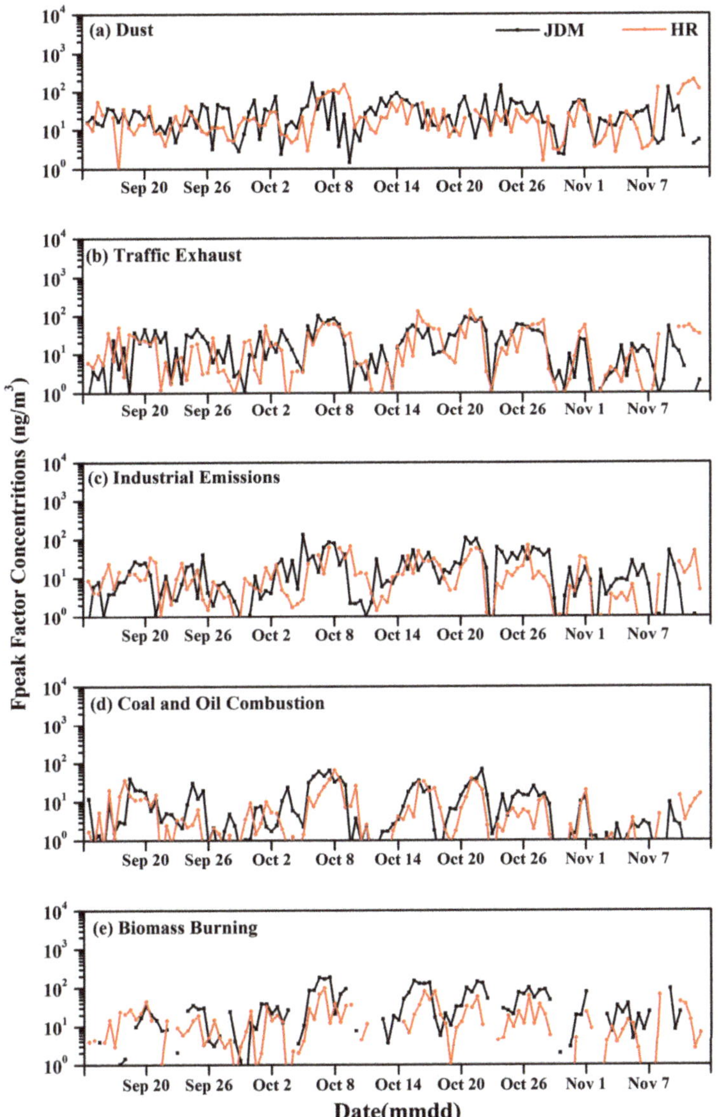

Figure 7. Time series of source contributions of the five factors ((**a**) dust, (**b**) traffic exhaust, (**c**) industrial sources, (**d**) coal and oil combustion and (**e**) biomass burning) to metallic elements in $PM_{2.5}$ resolved by PMF in Beijing urban site (JDM) and suburban site (HR).

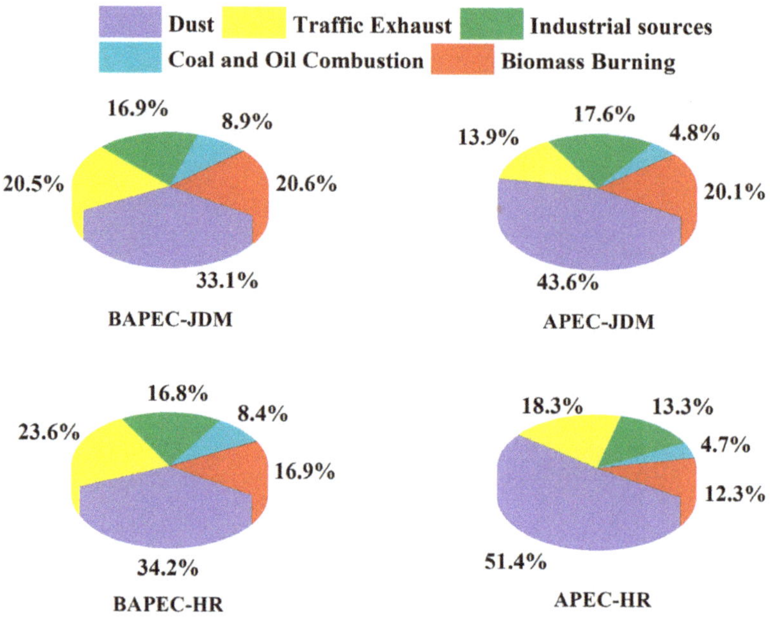

Figure 8. Source contribution of the five factor to metallic elements in $PM_{2.5}$ before and during the APEC summit in Beijing urban site (JDM) and suburban site (HR).

The first factor was identified as dust considering the abundance of the typical crustal elements (Mg, Al and Ca) [32–34]. Ca and Mg are the particular elements from cement and lime which mainly comes from construction dust. Besides, the Al and Mg take larger proportion among the yellow dust. Furthermore, the *EF*s of Ca and Mg were much lower than 10 in this study, indicating the nature of their origins. Therefore, this factor possibly mixed among construction dust and yellow dust. The average amounts of metallic elements from dust were approximately 162.0 ng/m^3 (33.1%) during the BAPEC period and 148.1 ng/m^3 (43.6%) during the APEC summit. Same monitoring values at HR were 95.5 ng/m^3 (34.2%) and 122.7 ng/m^3 (51.4%) respectively. The increased dust concentration at HR was possibly related to decreasing humidity levels (nearly 20%) during the APEC. The overall contribution from construction dust and soil dust was approximately 40% in 2007~2013 in Beijing [35], which is similar to the contribution rate from dust at HR in this study.

The second factor, which contained Zn, Cu and Pb, was identified as traffic exhaust. In the past, smelters and metallurgical industries were considered as the main sources of Cu and Zn in Beijing. However, metal smelting is no longer the main reason to makes Cu and Zn after relocating of the Capital Steel Corporation Limited in 2010. Zn was used in lubricant oil as an additive. Both Zn and Cu had been used in brake linings and tire manufacturing [36]. The friction between a brake pad and a brake plate will release Cu, and friction between the tire and the ground will release Zn. In China, Pb has been used within gasoline as an antidetonator. Both Zn and Pb have also been used as indicators of gasoline combustion in Hong Kong particle apportionment [37]. The concentration of Pb is low in this study which might be related with the promotion of unleaded gasoline. During the BAPEC period, the average amounts of metallic elements from traffic exhaust at JDM and HR were 100.2 ng/m^3 (20.5%) and 65.8 ng/m^3 (23.6%), respectively. Meanwhile, those during the APEC summit at JDM and HR were 47.3 ng/m^3 (13.9%) and 43.6 ng/m^3 (18.3%), respectively. These results are similar to the previous findings [38], that traffic emissions accounted for 22.9% and 21.7% before heating and during heating, respectively.

The third factor was identified as industrial sources due to its association with industrial elements (Cr, Mn and Fe). These components may be transported from the adjacent area Hebei Province, where a large number of highly polluting industries are located. Cr is a characteristic element of metallurgy and chemical dust. Both Cr and its compounds are widely used in metallurgy, electroplating, pigment, leather and other industries [39]. Previous study found that ferrous metallurgy could emit Mn [40]. Furthermore, both Fe and Mn are characteristic components of iron and steel industry emissions. Therefore, this factor could be identified as industrial sources. The average mass concentration of industrial sources decreased by 27.9% at JDM (from 83.0 ng/m^3 to 59.9 ng/m^3) and by 32.4% at HR (from 46.8 ng/m^3 to 31.6 ng/m^3) from the BAPEC period to during the APEC summit.

The fourth factor is identified as coal and oil combustion, which is suggested by the presence of V, Co, As and Cd. Coal is the dominant fuel in China. Its combustion is the predominant source of fine particulate matter over China [41]. The element As is characteristic of coal-fired dust [42], and coal combustion will also release some Co and Cd. V mainly originates from the combustion of petroleum (primarily heavy oil) and is easily discharged into the atmosphere in the form of particulate matter [43]. The contribution of coal and oil combustion in the BAPEC period were around 8% at both sites, while the contribution of this source decreased by 63.1% at JDM (from 43.7 ng/m^3 to 16.1 ng/m^3) and by 52.0% at HR (from 23.5 ng/m^3 to 11.3 ng/m^3) during APEC (Table 3). The contributions of coal and oil combustion in this study were much lower than the previous studies (11–38% in [18] and 26.1% in [44]), which may be attributed to the implementation of a coal-to-gas switch in recent years of Beijing.

The last factor was assigned to biomass burning because it contained a significant majority of K, which was a known biomass burning tracer [45]. K is a typical marker of biomass burning. Farming in Beijing's suburban districts has still been extensive in recent years. The farmers sometimes fertilize the soil and cooking by burning the crop remnants and the fallen leave in autumn and winter, which enhance the emission of K [46]. Besides, the several barbecue restaurants around the site JDM, which burning the carbon used in the barbecue may be enhance the biomass burning. During the BAPEC, the average mass concentrations of biomass combustion were 101.0 ng/m^3 (accounted for 20.6%) at JDM and 47.3 ng/m^3 (16.9%) at HR, while those same concentrations at JDM and HR were 68.4 ng/m^3 (20.1%) and 29.5 ng/m^3 (12.3%), respectively, during the APEC. These results are larger than the yearly average contribution of biomass burning in the year of 2010 (11.2%) [47], possibly because the sampling period of this study was consistent with the harvest period in Beijing and surrounding areas.

The mean contributions from the five sources decreased during the APEC summit. Generally, the five source contributions collectively diminished by 30.7% at JDM and 14.4% at HR during APEC relative to BAPEC, indicating that the air pollution regulations implemented during the APEC summit were effective, and the decline at JDM is more obvious than that at HR. These results are different to those from the prior study [14] stating that the contributions from seven sources (road dust, soil dust, traffic exhaust, secondary aerosols, industrial sources, biomass burning and residual oil combustion) collectively were reduced by 73.3% during the APEC summit at HR relative to the BAPEC period. It should be noted that secondary aerosol was not resolved in this study, which was the dominant factor attributed to the decline of $PM_{2.5}$ during the APEC summit [13]. In the present study, coal and oil combustion regulations were the most effective for reducing the pollution concentrations (JDM:63.1%; HR:52.0%), followed by measures to reduce traffic exhaust (52.8%), biomass burning (32.2%) and industrial sources (27.9%) at JDM and measures to reduce biomass burning (37.7%), traffic exhaust (33.7%) and industrial sources (32.4%) at HR. The variations in pollution from traffic sources during the BAPEC period and during the APEC summit at JDM were both greater than those at HR, while the variation in pollution from biomass combustion was the opposite.

4. Conclusions

Significant reductions in the $PM_{2.5}$ mass concentrations were observed both in the urban and suburban areas of Beijing during the APEC summit. The air quality evidently improved during APEC, during which period the mass concentrations at the urban site (JDM) and the suburban site (HR) were 48.4 $\mu g/m^3$ and 33.1 $\mu g/m^3$, which were 62.1% and 62.3% lower than those during the BAPEC period, respectively. Most of the trace metals (V, Cr, Mn, As, Cd and Pb) decreased more than 40% due to the emission regulations during APEC, while the crustal elements decreased considerably (4–45%). Relative to the daytime, trace metals increased during the nighttime at both sites before the APEC summit, but no significant difference was observed during the APEC summit, suggesting the suppressed emissions from anthropogenic activities. The concentration of the majority metallic elements increased with relatively aggravated pollution levels during the BAPEC period, and a significant increase ($p < 0.05$) was observed for K, Cr, Mn, Fe, Co, Cu, Zn, As, Ag, Cd and Pb when pollution levels changed from clean days to heavily polluted days. There was no significant difference for all the 15 elements when the pollution level changed from clean days to lightly polluted days. The *EF* results suggest trace metal elements are more easily enriched on pollution days. Five sources (dust, traffic exhaust, industrial sources, coal and oil combustion and biomass burning) were resolved using positive matrix factorization (PMF), which collectively decreased by 30.7% at the urban site and 14.4% at the suburban site during the APEC summit. Coal and oil combustion regulations were the most effective for reducing the trace element concentrations (urban site: 63.1%; suburban site: 52.0%), followed by measures to reduce traffic exhaust (52.8%) at the urban site and measures to reduce biomass burning (37.7%) at the suburban site. Our results suggest future control efforts for metallic elements in megacities like Beijing should prioritize coal and oil combustion as well as traffic emissions.

Supplementary Materials: The following are available online at http://www.mdpi.com/2073-4433/10/3/105/s1, Table S1: Concentrations of metallic elements in $PM_{2.5}$ in Beijing (ng/m^3), Table S2: PM2.5 Sources identification and source contributions in Beijing by PMF model in previous studies.

Author Contributions: Formal analysis, M.L.; Funding acquisition, Y.W.; Methodology, M.L.; investigation, X.H.; Y.X. and J.L.; data curation, B.H., J.C. and Y.Z.; Project administration, Y.W and Z.L.; Supervision, Z.L. and J.X; Writing—original draft, M.L.

Funding: This study was funded by the Ministry of Science and Technology of China (Grant nos. 2017YFC0210000), the National Natural Science Foundation of China (Grant nos. 41705110), Beijing Major Science and Technology Project (Z181100005418014), the National research program for key issues in air pollution control (DQGG0101) and the Strategic Priority Research Program of the Chinese Academy of Sciences (Grant nos. XDB05020200).

Acknowledgments: We would like to thank the staff in University of Chinese Academy of Sciences for the help in the sample collection.

Conflicts of Interest: The authors declare no conflict of interest. The funders had no role in the design of the study; in the collection, analyses, or interpretation of data; in the writing of the manuscript, or in the decision to publish the results.

References

1. Dockery, D.W.; Pope, C.A.; Xu, X.; Spengler, J.D.; Ware, J.H.; Fay, M.E.; Ferris, B.G., Jr.; Speizer, F.E. An association between air pollution and mortality in six U.S. cities. *New England J. Med.* **1993**, *329*, 1753–1759. [CrossRef] [PubMed]
2. Rd, P.C.; Dockery, D.W. Health effects of fine particulate air pollution: Lines that connect. *J. Air Waste Manage. Assoc.* **2006**, *56*, 709–742.
3. Wang, J.L.; Zhang, Y.H.; Shao, M.; Liu, X.L.; Zeng, L.M.; Cheng, C.L.; Xu, X.F. Quantitative relationship between visibility and mass concentration of $PM_{2.5}$ in Beijing. *J. Environ. Sci.* **2006**, *18*, 475–481.
4. Chow, J.C.; Watson, J.G.; Fujita, E.M.; Lu, Z.; Lawson, D.R.; Ashbaugh, L.L. Temporal and spatial variations of $PM_{2.5}$ and PM_{10} aerosol in the southern California air quality study. *Atmos Environ.* **1994**, *28*, 2061–2080. [CrossRef]
5. Donaldson, K.; Brown, D.; Clouter, A.; Duffin, R.; Macnee, W.; Renwick, L.; Tran, L.; Stone, V. The pulmonary toxicology of ultrafine particles. *J. Aerosol Med.* **2002**, *15*, 213–220. [CrossRef] [PubMed]

6. Prahalad, A.K.; Soukup, J.M.; Inmon, J.; Willis, R.; Ghio, A.J.; Becker, S.; Gallagher, J.E. Ambient air particles: Effects on cellular oxidant radical generation in relation to particulate elemental chemistry. *Toxicol. Appl. Pharmacol.* **1999**, *158*, 81–91. [CrossRef] [PubMed]
7. Gao, Y.; Nelson, E.D.; Field, M.P.; Ding, Q.; Li, H.; Sherrell, R.M.; Gigliotti, C.L.; Van Ry, D.A.; Glenn, T.R.; Eisenreich, S.J. Characterization of atmospheric trace elements on $PM_{2.5}$ particulate matter over the New York–New Jersey harbor estuary. *Atoms. Environ.* **2002**, *36*, 1077–1086. [CrossRef]
8. Pan, Y.P.; Wang, Y.S. Atmospheric wet and dry deposition of trace elements at 10 sites in Northern China. *Atmos. Chem. Phys.* **2015**, *15*, 951–972. [CrossRef]
9. Chen, Z.; Zhang, J.; Zhang, T.; Liu, W.; Liu, J. Haze observations by simultaneous lidar and WPS in Beijing before and during APEC, 2014. *Sci. China Chem.* **2015**, *58*, 1385–1392. [CrossRef]
10. Sheng, L.; Lu, K.; Ma, X.; Hu, J.K.; Song, Z.X.; Huang, S.X.; Zhang, J.P. The air quality of Beijing-Tianjin-Hebei regions around the Asia-Pacific Economic Cooperation (APEC) meetings. *Atmos. Pollut. Res.* **2015**, *6*, 1066–1072. [CrossRef]
11. Tang, G.; Zhu, X.; Hu, B.; Xin, J.; Wang, L.; Münkel, C.; Mao, G.; Wang, Y. Impact of emission controls on air quality in Beijing during APEC 2014: Lidar ceilometer observations. *Atoms. Chem. Phys.* **2015**, *15*, 12667–12680. [CrossRef]
12. Wen, W.; Cheng, S.; Chen, X.; Wang, G.; Li, S.; Wang, X.; Li, S.; Wang, X.; Liu, X. Impact of emission control on $PM_{2.5}$ and the chemical composition change in Beijing-Tianjin-Hebei during the APEC summit 2014. *Environ. Sci. Pollut. Res.* **2016**, *23*, 4509–4521. [CrossRef] [PubMed]
13. Liu, Z.; Hu, B.; Zhang, J.; Xin, J.; Wu, F.; Gao, W.; Wang, M.; Wang, Y. Characterization of fine particles during the 2014 Asia-Pacific economic cooperation summit: Number concentration, size distribution and sources. *Tellus B Chem Phys. Meteor.* **2017**, *69*. [CrossRef]
14. Wang, Y.; Zhang, Y.; Schauer, J.J.; De, F.B.; Guo, B.; Zhang, Y. Relative impact of emissions controls and meteorology on air pollution mitigation associated with the Asia-Pacific Economic Cooperation (APEC) conference in Beijing, China. *Sci. Total Environ.* **2016**, *571*, 1467–1476. [CrossRef] [PubMed]
15. Chang, Y.; Liu, X.; Deng, C.; Dore, A.J.; Zhuang, G. Source apportionment of atmospheric ammonia before, during, and after the 2014 APEC summit in Beijing using stable nitrogen isotope signatures. *Atmos. Chem. Phys.* **2016**, *16*, 1–26. [CrossRef]
16. Liacos, J.W.; Kam, W.; Delfino, R.J.; Schauer, J.J.; Sioutas, C. Characterization of organic, metal and trace element $PM_{2.5}$ species and derivation of freeway-based emission rates in Los Angeles, CA. *Sci. Total. Environ.* **2012**, *435–436*, 159–166. [CrossRef] [PubMed]
17. Yatkin, S.; Bayram, A. Source apportionment of PM_{10} and $PM_{2.5}$ using positive matrix factorization and chemical mass balance in Izmir, Turkey. *Sci. Total Environ.* **2008**, *390*, 109–123. [CrossRef] [PubMed]
18. Song, Y.; Tang, X.Y.; Xie, S.D.; Zhang, Y.H.; Wei, Y.J.; Zhang, M.S.; Zeng, L.; Lu, S. Source apportionment of $PM_{2.5}$ in Beijing in 2004. *J. Hazard. Mater.* **2007**, *146*, 124–130. [CrossRef] [PubMed]
19. Liu, Z.; Gao, W.; Yu, Y.; Hu, B.; Xin, J.; Sun, Y.; Wang, L.; Wang, G.; Bi, X.; Zhang, G.; et al. Characteristics of $PM_{2.5}$ mass concentrations and chemical species in urban and background areas of China: Emerging results from the CARE-China network. *Atmos. Chem. Phys.* **2018**, *18*, 8849–8871. [CrossRef]
20. Taylor, S.R.; Mclennan, S.M. The geochemical evolution of the continental crust. *Rev. Geophys.* **1995**, *33*, 293–301. [CrossRef]
21. Duce, R.A.; Hoffman, G.L.; Zoller, W.H. Atmospheric trace metals at remote northern and southern hemisphere sites: Pollution or natural? *Science* **1975**, *187*, 59–61. [CrossRef] [PubMed]
22. Paatero, P.; Tapper, U. Analysis of different modes of factor analysis as least squares fit problems. *Chemom. Intell. Lab. Syst.* **1993**, *18*, 183–194. [CrossRef]
23. Huang, S.; Arimoto, R.K. Testing and optimizing two factor-analysis techniques on aerosol at Narragansett, Rhode Island. *Atoms. Environ.* **1999**, *33*, 2169–2185. [CrossRef]
24. Polissar, A.V.; Hopke, P.K.; Paatero, P.; Malm, W.C.; Sisler, J.F. Atmospheric aerosol over Alaska: 2. Elemental composition and sources. *J. Geophy Res. Atmos.* **1998**, *103*, 19045–19057. [CrossRef]
25. Liu, Z.R.; Hu, B.; Liu, Q.; Sun, Y.; Wang, Y.S. Source apportionment of urban fine particle number concentration during summertime in Beijing. *Atoms. Environ.* **2014**, *96*, 359–369. [CrossRef]
26. Titos, G.; Lyamani, H.; Pandolfi, M.; Alastuey, A.; Alados-Arboledas, L. Identification of fine (PM1) and coarse (PM10-1) sources of particulate matter in an urban environment. *Atoms. Environ.* **2014**, *89*, 593–602. [CrossRef]

27. Moon, K.J.; Han, J.S.; Ghim, Y.S.; Kim, Y.J. Source apportionment of fine carbonaceous particles by positive matrix factorization at Gosan background site in East Asia. *Environ. Int.* **2008**, *34*, 654–664. [CrossRef] [PubMed]
28. Wang, L.; Liu, Z.; Sun, Y.; Ji, D.; Wang, Y. Long-range transport and regional sources of $PM_{2.5}$ in Beijing based on long-term observations from 2005 to 2010. *Atmos. Res.* **2015**, *157*, 37–48. [CrossRef]
29. Ram, K.; Sarin, M.M. Day-night variability of EC, OC, WSOC and inorganic ions in urban environment of Indo-Gangetic Plain: Implications to secondary aerosol formation. *Atoms. Environ.* **2011**, *45*, 460–468. [CrossRef]
30. Qiao, B.W.; Liu, Z.R.; Hu, B.; Liu, J.Y.; Pang, N.N.; Wu, F.K.; Xu, Z.J.; Wang, Y.S. Concentration Characteristics and Sources of Trace Metals in $PM_{2.5}$ During Wintertime in Beijing. *Environ Sci.* **2017**, *38*, 876–883. (in Chinese).
31. Yang, Y.J.; Wang, Y.S.; Huang, W.W.; Hu, B.; Wen, T.X. Size Distributions and Elemental Compositions of Particulate Matter on Clears Hazy and Foggy days in Beijing, China. *Adv. Atmos. Sci.* **2010**, *27*, 663–675. [CrossRef]
32. Watson, J.G.; Chow, J.C. Source characterization of major emission sources in the imperial and Mexicali Valleys along the US/Mexico border. *Sci. Total Environ.* **2001**, *276*, 33–47. [CrossRef]
33. Zhao, W.; Hopke, P.K. Source apportionment for ambient particles in the San Gorgonio wilderness. *Atoms. Environ.* **2004**, *38*, 5901–5910. [CrossRef]
34. Kim, E.; Hopke, P.K.; Edgerton, E.S. Improving source identification of Atlanta aerosol using temperature resolve carbon factions in positive matrix factorization. *Atoms. Environ.* **2004**, *38*, 3349–3362. [CrossRef]
35. Jin, X.C.; Zhang, G.Y.; Xiao, C.J.; Huang, D.H.; Yuan, G.J.; Yao, Y.G.; Wang, X.H.; Hua, L.; Wang, P.S.; Ni, B. Source Apportionment of $PM_{2.5}$ in Xinzhen, Beijing Using PIXE and XRF. *Energy Sci. Technol.* **2014**, *48*, 1325–1330. (in Chinese).
36. Thorpe, A.; Harrison, R.M. Sources and properties of non-exhaust particulate matter from road traffic: A review. *Sci. Total Environ.* **2008**, *400*, 270–282. [CrossRef] [PubMed]
37. Lee, E.; Chan, C.K.; Paatero, P. Application of positive matrix factorization in source apportionment of particulate pollutants in Hong Kong. *Atoms. Environ.* **1999**, *33*, 3201–3212. [CrossRef]
38. Yang, H.; Chen, J.; Wen, J.; Tian, H.; Liu, X. Composition and sources of $PM_{2.5}$ around the heating periods of 2013 and 2014 in Beijing: Implications for efficient mitigation measures. *Atoms. Environ.* **2016**, *124*, 378–386. [CrossRef]
39. Dall'Osto, M.; Querol, X.; Amato, F.; Karanasiou, A. Hourly elemental concentrations in $PM_{2.5}$ aerosols sampled simultaneously at urban background and road site during SAPUSS-diurnal variations and PMF receptor modelling. *Atmos. Chem. Phys.* **2013**, *13*, 4375–4392.
40. Querol, X.; Zhuang, X.; Alastuey, A.; Viana, M.; Lv, W.; Wang, Y.; Lopez, A.; Zhu, Z.; Wei, H.; Xu, S. Speciation and sources of atmospheric aerosols in a highly industrialised emerging mega-city in central China. *J. Environ. Monitor.* **2006**, *8*, 1049–1059. [CrossRef]
41. Yao, Q.; Li, S.Q.; Xu, H.W.; Zhuo, J.K.; Song, Q. Reprint of: Studies on formation and control of combustion particulate matter in China: A review. *Energy* **2009**, *35*, 4480–4493. [CrossRef]
42. Hien, P.D.; Binh, N.T.; Truong, Y.; Ngo, N.T.; Sieu, L.N. Comparative receptor modelling study of TSP, PM_2 and PM_{2-10} in Ho Chi Minh City. *Atoms. Environ.* **2001**, *35*, 2669–2678. [CrossRef]
43. Wang, G.; Oldfield, F.; Xia, D.; Chen, F.; Liu, X.; Zhang, W. Magnetic properties and correlation with heavy metals in urban street dust: A case study from the city of Lanzhou, China. *Atoms. Environ.* **2012**, *46*, 289–298. [CrossRef]
44. Huang, R.J.; Zhang, Y.; Bozzetti, C.; Ho, K.F.; Cao, J.J.; Han, Y.; Daellenbach, K.R.; Slowik, J.G.; Platt, S.M.; Canonaco, F.; et al. High secondary aerosol contribution to particulate pollution during haze events in China. *Nature* **2014**, *514*, 218–222. [CrossRef] [PubMed]
45. Cachier, H.; Ducret, J. Influence of biomass burning on equatorial African rains. *Nature* **1991**, *352*, 228–230. [CrossRef]
46. Duan, F.; Yu, L.T.; Cachier, H. Identification and estimate of biomass burning contribution to the urban aerosol organic carbon concentrations in Beijing. *Atoms. Environ.* **2004**, *38*, 1275–1282. [CrossRef]
47. Yu, L.D.; Wang, G.F.; Zhang, R.J.; Zhang, L.M.; Song, Yu.; Wu, B.B. Characterization and Source Apportionment of $PM_{2.5}$ in an Urban Environment in Beijing. *Aerosol. Air. Qual. Res.* **2013**, *13*, 574–583. [CrossRef]

Article

Levels and Sources of Atmospheric Particle-Bound Mercury in Atmospheric Particulate Matter (PM_{10}) at Several Sites of an Atlantic Coastal European Region

Jorge Moreda-Piñeiro [1,2,3,*], Adrián Rodríguez-Cabo [1], María Fernández-Amado [1,2,3], María Piñeiro-Iglesias [1,2,3], Soledad Muniategui-Lorenzo [1,2,3] and Purificación López-Mahía [1,2,3]

1 Grupo Química Analítica Aplicada (QANAP), Department of Chemistry, Faculty of Sciences, University of A Coruña, Campus de A Coruña, s/n. 15071–A Coruña, Spain; adrian.cabo@udc.es (A.R.-C.); maria.fernandez.amado@udc.es (M.F.-A.); maria.pineiro.iglesias@udc.es (M.P.-I.); smuniat@udc.es (S.M.-L.); purmahia@udc.es (P.L.-M.)

2 University Institute of Research in Environmental Studies (IUMA), University of A Coruña, Pazo de Lóngora, 15179 Liáns, Oleiros–A Coruña, Spain

3 Centro de Investigaciones Científicas Avanzadas (CICA), University of A Coruña, Campus de Elviña, s/n. 15008–A Coruña, Spain

* Correspondence: jorge.moreda@udc.es; Tel.: +34-981-167000

Received: 25 November 2019; Accepted: 23 December 2019; Published: 27 December 2019

Abstract: Atmospheric particle-bound mercury (PHg) quantification, at a pg m^{-3} level, has been assessed in particulate matter samples (PM_{10}) at several sites (industrial, urban and sub-urban sites) of Atlantic coastal European region during 13 months by using a direct thermo-desorption method. Analytical method validation was assessed using 1648a and ERM CZ120 reference materials. The limits of detection and quantification were 0.25 pg m^{-3} and 0.43 pg m^{-3}, respectively. Repeatability of the method was generally below 12.6%. PHg concentrations varied between 1.5–30.8, 1.5–75.3 and 2.27–33.7 pg m^{-3} at urban, sub-urban and industrial sites, respectively. PHg concentration varied from 7.2 pg m^{-3} (urban site) to 16.3 pg m^{-3} (suburban site) during winter season, while PHg concentrations varied from 9.9 pg m^{-3} (urban site) to 19.3 pg m^{-3} (suburban site) during the summer. Other trace elements, major ions, black carbon (BC) and UV-absorbing particulate matter (UV PM) was also assessed at several sites. Average concentrations for trace metals (Al, As, Bi, Cd, Cr, Cu, Fe, Mn, Ni, Pb, Sb, Si, Sr, V and Zn) ranged from 0.08 ng m^{-3} (Bi) at suburban site to 1.11 µg m^{-3} (Fe) at industrial site. Average concentrations for major ions (including Na^+, K^+, Ca^{2+}, $NH_4{}^+$, Mg^{2+}, Cl^-, $NO_3{}^-$ and $SO_4{}^{2-}$) ranged from 200 ng m^{-3} (K^+) to 5332 ng m^{-3} ($SO_4{}^{2-}$) at urban site, 166 ng m^{-3} (Mg^{2+}) to 4425 ng m^{-3} ($SO_4{}^{2-}$) at suburban site and 592 ng m^{-3} (K^+) to 5853 ng m^{-3} (Cl^-) at industrial site. Results of univariate analysis and principal component analysis (PCA) suggested crustal, marine and anthropogenic sources of PHg in PM_{10} at several sites studied. Toxicity prediction of PHg, by using hazard quotient, suggested no non-carcinogenic risk for adults.

Keywords: atmospheric particle-bound mercury; atmospheric particulate matter; sources contributions; Atlantic coastal European region; toxicity prediction

---

1. Introduction

Inhalation of atmospheric particulate matter (APM) represents a significant exposure pathway to humans. Several epidemiological studies have shown that chronic environmental exposure to APM has been associated with specific negative health outcomes (decreased pulmonary and renal function; lung cancer; damage to DNA; and cardiovascular, reproductive and endocrine alterations) [1]. The World Health Organization estimates that around six million people in the world die annually due to the effects of atmospheric pollution (indoor and outdoor) [2].

Mercury is a metal without any physiological demand in humans, with high toxicity (it affects the human central nervous system), long-distance transport and strong bioaccumulation tendency in the food chain, which poses a global concern and a great threat to human health, wildlife and environment [3]. Hg and its species are therefore included in priority lists of toxic compounds by several international agreements dealing with environmental protection and international programs to reduce mercury emissions [4–7]. Mercury is emitted into the atmosphere, as gaseous elemental mercury Hg^0 (GEM) and particle-bound mercury (PHg), from both natural (volcanic eruptions, sandstorms, crustal dust and rock weathering, evaporation from water surfaces, geothermal vents and forest fires) and anthropogenic (coal combustion, waste incineration and cement, chlor-alkali, nonferrous metal production) sources [8,9]. GEM is the most predominant form of total gaseous mercury, over 95%, into the atmosphere [10]. GEM is transformed into the atmosphere, via redox chemistry and homogeneous reactions, to gaseous oxidized mercury (GOM), which can be converted to PHg upon adsorption/absorption on aerosol surfaces [11–13]. PHg, which accounts for less than 10% of the total atmospheric Hg [14] can then undergo both dry and wet deposition [13].

In the last decades, several studies have focused on the determination of PHg levels at different sites of the Europe [15–21], but data on levels of PHg in PM_{10} at Atlantic coastal European regions studies are scarce [22–26]. Measurements of PHg in APM generally fell in the range of few pg m^{-3} to ng m^{-3} at rural and urban areas; thus, high and selective techniques, such as inductively coupled plasma mass spectrometry [27], cold vapour atomic absorption/fluorescence spectrometry [28,29], and differential pulse anodic stripping voltammetry [30], are required. These methodologies cannot be considered as environmentally friendly processes due to the use of toxic reagents/acids at high concentrations during the sample pre-treatment. Recently, a more effective, high sensitivity, fast, environmentally friendly and simple direct atomic absorption spectrometry methods followed by thermal decomposition of sample in an oxygen-rich atmosphere and concentration of mercury vapour on an amalgamator have been proposed [31,32]. This method is more suitable for PHg determination due to reduction of Hg losses and the possibility of sample contamination.

In this study we quantified mercury concentrations in PM_{10} (mean APM which passes through a size-selective inlet with a 50% efficiency cut-off at 10 μm aerodynamic diameter) at pg m^{-3} levels by direct solid sampling atomic absorption spectrometry. PHg levels were compared with other studies carried out in Europe and in Atlantic coastal European region. PHg seasonal variability and their relationships with APM sources at several sites (industrial, urban and sub-urban sites) of the Atlantic coastal European region during 13 months were also examined. Finally, due to the high toxicity of mercury and scarce information about mercury human health-risk assessment via inhalation several hazard indexes such as average daily intake (ADI), and hazard quotient (HQ) were assessed.

2. Materials and Methods

2.1. Details and Description of the Study Areas

A Coruña is an Atlantic coastal city in the northwest of Spain with a quarter of a million inhabitants (Figure 1). Due to its proximity to the sea, the sea salt content of particulate matter is also important. Climate of zone is humid oceanic, with abundant rainfall and prevailing winds from the north in summer and south in winter [33]. The main anthropogenic sources in this area are the emissions from traffic and domestic activities, industrial emissions and biomass burning [33]. Samples were simultaneously collected from May 2009 to May 2010, at urban, sub-urban and industrial sites. Urban site (A Coruña, US, Spain), located inside the downtown (coordinates: 43°22′04″ N 08°25′08″ W) at 5 m above the sea level. Sub-urban site (Liáns, SS, Spain), a residential area near A Coruña city (coordinates: 43°20′12″ N 08°21′09″ W) at 120 m above the sea level (Moreda-Piñeiro et al., 2014).

Industrial site (Sabón, IS), an industrial area near to A Coruña (coordinates: 43°19′36″ N 08°30′0.2″ W) at 62 m above the sea level.

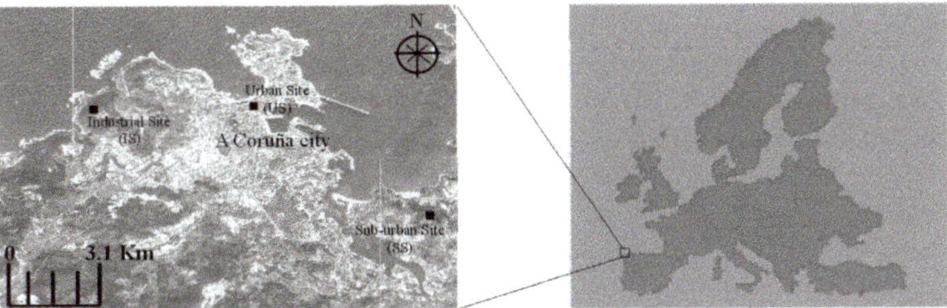

Figure 1. Sampling sites at the A Coruña City.

2.2. Atmospheric Particulate Matter Sample Collection

PM samples were collected using high-volume samplers (Graseby-Andersen, Gibsonville, NC, USA) with a PM_{10} head. The samples were taken in 24 h intervals (from 11:00 a.m. to 11:00 a.m. next day) during the whole measurement period. Sampler meets the requirements of UNE EN 12,341 European Norm (European standard EN12341, 2015) [34]. PM_{10} collection was conducted on QF20 quartz fiber filters (Schleischer&Schuell, Dassel, Germany). Before sampling, the filters were pre-heated at 450 °C for 1 h to remove possible mercury and other trace metal contamination. Before and after sampling, filters were stabilized at 20 ± 1 °C and relative humidity of 50 ± 5% for 48 h, for mass determination by means of a microbalance (Sartorius Genius, Goettingen, Germany) with an accuracy of 0.01 mg [34]. Afterwards sampling, PM_{10} samples were wrapped in aluminum foil, were sealed in clean polyethylene zipper bags and then stored in a freezer (−18 °C) until further analysis to avoid the losses of mercury from the filter. A total of 123 samples were collected covering thirteen months (44, 38 and 41 samples collected at US, SS and IS sites, respectively). In general, 1–2 samples per week were collected covering sampling period. Finally, field blanks were analysed following the same procedure used for the samples.

2.3. Cleaning Procedures

All plastic ware and glassware were washed with ultrapure water of 18 MΩ cm resistance (Milli-Q water purification system, Millipore, Bedford, MA, USA). Then, plastic ware and glassware were soaked for 48 h with 10% (v/v) nitric acid (ultraclean nitric acid 69–70%, JT Baker, Phillipsburg, PA, USA), and rinsed several times with ultrapure water before use. Sampling boats were pre-heated at 600 °C for 15 min to remove possible mercury contamination. After collection, sample manipulation and analysis were carried out into a class-100 clean room.

2.4. Mercury Quantification by DMA-80 Direct Mercury Analyser

Mercury measurements were performed on the Direct Mercury Analyser DMA-80 spectrometer (Milestone Srl, Sorisole, Italy) by the AAS technique. DMA-80 is a single-purpose atomic absorption spectrometer for determination of mercury traces in various solids and liquids without sample pre-treatment/pre-concentration. Three circular pieces (3 × 9.42 cm^2), cut from each PM_{10} filter, were placed into a sampling boat and transferred to DMA-80. The methodology is based on a thermal decomposition of the sample in an oxygen-rich atmosphere (99.5%) for 60 s at 750 °C; transportation of released gasses (via an oxygen carrier gas) through specific catalytic compounds (for interfering impurities removal) to an Au-amalgamator; and collection of the Hg vapor on an tamalgamator. After a pre-defined time the amalgamator was heated up to 800 °C. The released Hg was transported to to the detection system, which contains the Hg-specific lamp (253.7 nm) for mercury quantification. Before the next analysis, the system was cleaned for 45 s to avoid any memory effects.

For quantification, a mercury stock standard solution 1000 mg L^{-1} (Merck, Poole, Dorset, UK) as Hg^{2+} in dilute nitric acid (Baker) was used to prepare aqueous calibration solutions in the working range of 0–10 ng. The polynomial of degree two curve was: $Signal_{Hg} = -0.006\,[Hg]^2 + 0.1641\,[Hg] + 0.0042\,\left(R^2 = 0.9998\right)$.

Analytical recovery studies were performed to procedure validation. Three circular pieces from a PM_{10} filter (9.42 cm^2) were spiked with different amounts of mercury at four levels: 0.25 and 0.5 ng (low levels), 1.0 ng (intermediate level) and 2.0 ng (high level). Each concentration level was performed five times, and also, the un-spiked filter sample was also subjected to the mercury quantification (five times). Analytical recoveries within the 87–110% range were obtained, and good accuracy has been proved. Also, accuracy of the method was assessed by analysing SRM 1648a urban particulate matter (National Institute of Standards and Technology, Gaithersburg, MD, USA). Reference material was analysed ten times by using the described methodology. Concentrations found (1.25 ± 0.09 mg Kg^{-1}) is in good agreement with the certified value (1.323 ± 0.064 mg Kg^{-1}) after statistical evaluation by applying a *t*-test at 95% confidence level for nine degrees of freedom. tcal value (2.21) is lower than the ttab value of 2.26 Therefore, mercury in PM_{10} can directly be determined after thermal decomposition, collection on an amalgamator and AAS detection ensuring the accuracy of analysis. Finally, ERM CZ120 fine dust (like PM_{10}) (European CommissionJoint Research Centre Institute for Reference Materials and Measurements (IRMM), Geel, Belgium) reference material was also analysed, concentrations found was 0.22 ± 0.02 mg Kg^{-1}. Unfortunately, no certified/informative or reported values by other authors were available for this reference material.

The limit of detection (LOD) and the limit of quantification (LOQ), based on the 3σ and 10σ criterion (σ, the standard deviation of background signal), were calculated. Keeping in mind the filter surface (9.42 cm^2) and the air volume taken through, the LODs and LOQs were 0.25 and 0.43 pg m^{-3}, respectively; where it can be seen that the values are low enough to perform mercury quantification in PM_{10} samples. The repeatability of the procedure was obtained by subjecting a PM_{10} sample 11 times to the proposed procedure. Low RSD values (12.6%) was achieved.

2.5. Statistical Treatment of Data

To evaluate the analytical data, Univariate Analysis, Correlation Analysis and Principal Component Analysis were performed with Statgraphics version 7.0 (StatgraphicsGraphics Corporation, SC, USA).

3. Results and Discussion

3.1. Atmospheric Particle-Bound Mercury Concentration in PM_{10}

Average PHg concentrations at three sites (US, SS and IS), along with statistical results (RSD, maximum and minimum concentration, and range) are summarized in Table 1. High PHg content in PM_{10} concentrations were observed at SS (PHg concentrations varied between 1.5 and 75.3 pg m^{-3}), while low values were achieved at US (PHg concentrations varied between 1.5 and 30.8 pg m^{-3}). The biomass burning associated to agricultural activities and wood combustion for heating purposes at SS could be explain the high PHg concentration at SS (see the next section). A first attempt to find tendencies consisted of comparing the averages and SDs of PHg from three sites, and results from ANOVA, showed statistically significant differences (95.0% confidence level) between US and SS sites (the *p*-values of the F-test is lower than 0.05).

PHg concentrations found (Table 1) are in the concentration range reported for several European sites [15,17–21]. PHg concentration in total particulate matter varied from 5 to 200 pg m^{-3} at several sites of Northern Europe and in the Mediterranean region [21], from 0.11 to 1.05 pg m^{-3} in rural air in Southern Poland [20] and from 3.9 to 20.3 pg m^{-3} at Göteborg [19]. Lewandowska et al. reported PHg concentrations in the range of 0.6 to 32.9 pg m^{-3} in small aerosols (PM_1) at an urbanized coastal zone of the Gulf of Gdansk (southern Baltic) [15]. PHg concentrations in the range of 2 to 142 pg m^{-3} were reported in coarse and fine particles ($PM_{0.4\text{–}2.0}$) at the same site by Beldowska et al. [18]. Low concentrations have been also reported at urban (7.3–22.6 pg m^{-3}) and forest (2.4–20.8 pg m^{-3})

sites of Poland in coarse ($PM_{2.2}$) and fine ($PM_{0.7}$) particles [17]. Similar PHg levels (61 and 66 ng m^{-3} in $PM_{2.5}$ and $PM_{2.5-10}$, respectively) have been reported at a sub-urban area at the Western European coast (Portugal) [26].

Table 1. Average, SD, range and minimum and maximum values of PHg in PM_{10} samples at urban site (US), suburban site (SS) and industrial site (IS).

Site	Average ± SD (pg m^{-3})	Max (pg m^{-3})	Min (pg m^{-3})	Range (pg m^{-3})	N
			Whole period		
US	8.5 ± 7.3	30.8	1.5	29.4	44
SS	18.0 ± 19.8	75.3	1.5	73.8	38
IS	11.4 ± 7.6	33.7	2.3	31.4	41
			Summer season		
US	7.3 ± 5.9	25.7	1.5	24.2	22
SS	16.4 ± 19.0	61.9	1.5	60.4	17
IS	7.5 ± 2.8	13.0	2.3	10.7	18
			Winter season		
US	9.9 ± 8.4	30.8	1.5	29.3	22
SS	19.3 ± 20.4	75.3	1.5	73.8	21
IS	14.4 ± 8.6	33.7	2.8	30.9	23

On the contrary, high PHg values have been published for several Atlantic European sites. Arruti et al. reported PHg concentrations of <0.7–1.4 ng m^{-3} in PM_{10} samples and 0.8–0.9 ng m^{-3} in $PM_{2.5}$ samples at several urban, suburban and industrial sites of the Cantabria region (Northern Spain) [22,23]. Frietas et al. reported PHg levels in the range of 0.14 to 1.5 and 0.07 to 2.3 ng m^{-3} in PM_{10} and $PM_{2.5}$ samples, respectively in several inland sites of Portugal [25]. Finally, PHg levels around 0.9 ng m^{-3} in $PM_{2.5}$ and $PM_{2.5-10}$ samples were reported at industrial sites of Portugal by Farinha et al. [24]. The analysis of samples from industrial and contaminated sites and the use of analytical techniques such NAA could explain the high PHg concentrations reported.

A PHg variation among PM_{10} samples was achieved during the sampling period at several sites, RSD higher than 66% (Table 1), which reflects inherent heterogeneity of the atmospheric particles (marine, crustal and anthropogenic/industrial contributions). The variation of monthly mean PHg concentration (Figure 2) in PM_{10} at several sites varied largely from month to month. Descriptive statistics of PHg during the winter and summer seasons are summarised in Table 1. During the winter season (January–March and October–December), PHg concentration varied from 1.5 to 30.8, 1.5 to 75.3 and 2.8 to 33.7 pg m^{-3} at US, SS and IS, respectively (Table 1); with an average of 9.9 ± 8.4, 19.3 ± 20.4 and 14.4 ± 8.6 pg m^{-3} for US, SS and IS, respectively. On the contrary, concentrations of PHg varied from 1.5 to 25.7, 1.5 to 61.9 and 2.3 to 13.1 pg m^{-3} at US, SS and IS, respectively; with an average of 7.3 ± 5.9, 16.4 ± 19.0 and 7.5 ± 2.8 pg m^{-3} for US, SS and IS, respectively (Table 1) during the summer period (April–September). Although, high mean PHg concentrations were obtained during the winter season (November and December 2014 and January, February and March 2015) at all sites (Table 1, Figure 2); after applying ANOVA (Table 2), statistically significant differences (95.0% confidence level) were found between summer and winter seasons only at IS (the p-value of the F-test is lower than 0.05). Without neglecting the influence of the meteorological parameters (humidity, ambient temperature, precipitation, wind speed and direction), photochemical transformation and gas-particulate phase partitioning of atmospheric Hg, the high concentration of PHg during the winter may be due to the increased emissions from anthropogenic sources [30].

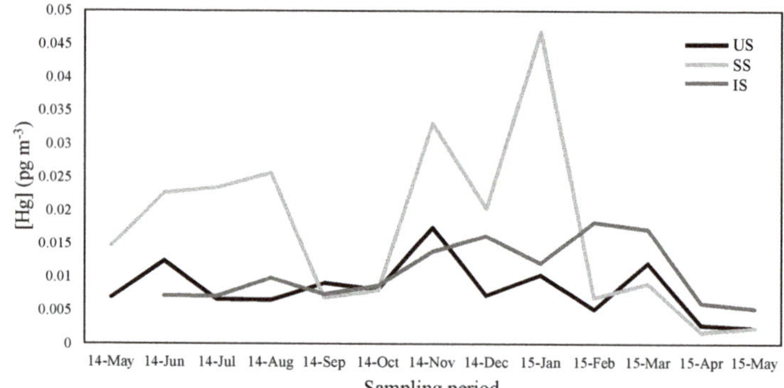

Figure 2. Monthly variation of PHg concentration during the study period at urban (US), suburban (SS) and industrial (IS) sites.

Table 2. Analysis of variance of PHg concentration for winter and summer seasons at urban (US), suburban (SS) and industrial (IS) sites.

Site	Groups	Sum of Squares	Degree of Freedom	Mean Square	F-Ratio	p-Value
US	Between groups	73.79	1	73.79	1.41	0.2420
	Within groups	2200.1	42	52.38		
SS	Between groups	79.64	1	79.64	0.20	0.6547
	Within groups	14090.8	36	391.4		
IS	Between groups	489.90	1	489.90	10.77	0.0022
	Within groups	1774.0	39	45.49		

3.2. Sources of PHg in PM_{10}

A study on the relations between Hg content (Table 1) and major ions, trace metal, equivalent black carbon (eBC) and UV-absorbing particulate matter (UVPM) concentrations (Tables S1 and S2) in PM_{10}, and consequently the PHg sources, was assessed. Univariate and PCA analysis were assessed for samples collected at US (whole period $N = 44$, summer season $N = 22$ and winter season $N = 22$), SS (whole period $N = 38$, summer season $N = 17$ and winter season $N = 21$) and IS (whole period $N = 41$, summer season $N = 182$ and winter season $N = 23$). Procedures for major ions and trace metals extraction and quantification are summarized in the Supporting Information section; details for eBC and UVPM are also shown in this section.

3.2.1. Univariate Analysis

Spearman coefficients and p-values (Table 3) between PHg and some ions and trace metals for samples collected at US during the whole measuring period were as follows: NO_3^- (0.5889, p-value = 0.0022); Cr (0.6688, p-value = 0.0005); Pb (0.6037, p-value = 0.0017); and Sb (0.3979, p-value = 0.0387). Also, good correlation (p-value lower than 0.05) were found for PHg and eBC (0.4948, p-value = 0.0101) and UVPM (0.4226, p-value = 0.0281). These results indicate the anthropogenic origin of PHg, mainly associated with combustion of fossil fuels and road traffic. Similar conclusions were reached for samples collected during winter and summer seasons at US.

Good correlations were found for PHg and NO_3^- (0.6147, p-value = 0.0173 and 0.5874, p-value = 0.0514 for summer and winter seasons, respectively); for PHg and Cr (0.4742, p-value = 0.0562 and 0.7018, p-value = 0.0190 for summer and winter seasons, respectively); PHg and Pb (0.5647, p-value = 0.0519 and 0.7413, p-value = 0.0140 for summer and winter seasons, respectively); and PHg and eBC (0.5353, p-value = 0.0118 and 0.4216, p-value = 0.0186 for summer and winter seasons, respectively).

Table 3. Spearman rank correlation coefficients and *p*-values (in brackets) for PHg and major ions, metals, eBC and UVPM at urban site US (whole period, summer season, winter season), suburban site SS (whole period, summer season, winter season) and industrial site IS (whole period, summer season, winter season) at the probability of 95%. Bolded numbers means statistically significant correlations.

	US			SS			IS		
	Summer + Winter	**Summer**	**Winter**	**Summer + Winter**	**Summer**	**Winter**	**Summer + Winter**	**Summer**	**Winter**
PM_{10} mass	0.1357 (0.4806)	−0.0.353 (0.893)	0.1818 (0.5465)	0.2773 (0.1059)	−0.0441 (0.8599)	0.4510 (0.0630)	0.10003 (0.5706)	0.1490 (0.5372)	−0.3679 (0.1687)
NH_4^+	−0.0126 (0.9478)	−0.0588 (0.8199)	0.1399 (0.6427)	0.2199 (0.1998)	0.2725 (0.2765)	0.1125 (0.6428)	0.3390 (0.0551)	−0.0258 (0.9153)	0.4808 (0.0720)
K^+	−0.1232 (0.5222)	−0.0353 (0.8913)	−0.3986 (0.1862)	0.4380 (0.0653)	−0.1533 (0.5398)	**0.4778 (0.0488)**	0.0889 (0.6150)	0.3106 (0.2003)	−0.0107 (0.9680)
Na^+	−0.3678 (0.0560)	−0.3441 (0.1826)	−0.5224 (0.0690)	**−0.3930 (0.0219)**	**−0.5196 (0.0377)**	−0.2528 (0.2972)	−0.2390 (0.1764)	−0.4299 (0.0818)	−0.3143 (0.2396)
Ca^{2+}	0.2063 (0.2836)	0.3029 (0.2407)	−0.0909 (0.7630)	0.2832 (0.0987)	0.0613 (0.8064)	**0.5955 (0.0141)**	−0.0692 (0.6955)	−0.0712 (0.7691)	0.0536 (0.8411)
Mg^{2+}	−0.3607 (0.0609)	−0.3441 (0.1826)	−0.5315 (0.0780)	−0.2625 (0.1259)	−0.5000 (0.8064)	−0.0093 (0.9694)	−0.1270 (0.4725)	0.2590 (0.2855)	−0.2714 (0.3098)
Cl^-	−0.2507 (0.1927)	−0.2206 (0.3929)	−0.3427 (0.2558)	**−0.5532 (0.0376)**	−0.4843 (0.0554)	0.0114 (0.9627)	−0.1848 (0.2958)	0.3228 (0.1126)	−0.4071 (0.1277)
SO_4^{2-}	0.0980 (0.6107)	0.0941 (0.7155)	0.2168 (0.4721)	**0.4476 (0.0091)**	**0.6201 (0.0131)**	0.3416 (0.1590)	0.2721 (0.1238)	0.1641 (0.4987)	0.5179 (0.0527)
NO_3^-	**0.5889 (0.0022)**	**0.6147 (0.0173)**	0.5874 (0.0514)	0.1745 (0.3038)	0.2940 (0.2470)	0.1455 (0.5485)	0.1374 (0.4371)	0.1187 (0.6246)	0.3000 (0.2617)
Al	0.0082 (0.9659)	−0.1987 (0.4416)	0.3123 (0.3003)	0.1843 (0.2825)	0.0650 (0.7949)	**0.5041 (0.0495)**	**0.3457 (0.0505)**	−0.0506 (0.8348)	**0.6214 (0.0201)**
As	−0.0438 (0.8199)	−0.3761 (0.1452)	0.4273 (0.1664)	0.4715 (0.0600)	0.4012 (0.1085)	**0.6144 (0.0113)**	0.2669 (0.0930)	0.1166 (0.6306)	0.3330 (0.2127)
Bi	−0.0827 (0.6674)	−0.4418 (0.0870)	0.3726 (0.2166)	0.3305 (0.0539)	0.3640 (0.1454)	0.2884 (0.2344)	**0.3538 (0.0454)**	0.1770 (0.4665)	**0.5332 (0.0460)**
Cd	0.0398 (0.8360)	−0.1624 (0.5295)	0.3410 (0.2581)	−0.0131 (0.8552)	−0.2730 (0.2731)	0.1034 (0.6700)	**0.4032 (0.0226)**	0.1613 (0.5059)	0.4861 (0.0689)
Cr	**0.6688 (0.0005)**	0.4742 (0.0562)	**0.7018 (0.0190)**	**0.3467 (0.0432)**	**0.4945 (0.0479)**	0.3038 (0.2104)	−0.0351 (0.8425)	−0.1043 (0.6670)	−0.0717 (0.7885)
Cu	0.3637 (0.0590)	0.3824 (0.1386)	0.2168 (0.4721)	0.0014 (0.9935)	0.1693 (0.4982)	−0.2189 (0.3668)	0.1765 (0.3181)	−0.0609 (0.8018)	0.3143 (0.2396)
Fe	**0.4155 (0.0308)**	0.3735 (0.1480)	0.3930 (0.1924)	0.3155 (0.0658)	0.3004 (0.2295)	0.5346 (0.0532)	0.4373 (0.0514)	0.1168 (0.6301)	**0.5964 (0.0380)**
Mn	0.1796 (0.3506)	0.0154 (0.9526)	0.3556 (0.2383)	−0.0302 (0.8602)	0.1357 (0.5873)	−0.1082 (0.6555)	0.2653 (0.1334)	−0.04303 (0.8681)	0.2906 (0.2768)
Ni	−0.1692 (0.3792)	**−0.5133 (0.0468)**	0.2168 (0.4721)	0.2757 (0.1079)	0.2876 (0.2499)	0.2851 (0.2398)	0.3061 (0.0834)	0.0918 (0.7049)	0.3682 (0.1683)
Pb	**0.6037 (0.0017)**	0.5647 (0.0519)	**0.7413 (0.0140)**	0.2448 (0.1534)	0.2451 (0.3264)	0.2528 (0.2972)	0.4106 (0.0514)	−0.0031 (0.9898)	0.1107 (0.6787)
Sb	**0.3979 (0.0387)**	0.2647 (0.3053)	0.4266 (0.1571)	0.3232 (0.0549)	0.2770 (0.2679)	0.3540 (0.1444)	**0.4405 (0.0127)**	0.2322 (0.3384)	**0.6500 (0.0150)**
Si	0.0077 (0.9683)	−0.1853 (0.4730)	0.3047 (0.3122)	0.1840 (0.2834)	0.0650 (0.7949)	**0.5060 (0.0441)**	**0.3522 (0.0463)**	−0.0527 (0.8278)	**0.3214 (0.0201)**
Sr	0.0071 (0.9705)	0.1941 (0.4522)	−0.2448 (0.4169)	−0.2459 (0.1516)	−0.1642 (0.5113)	−0.2797 (0.2489)	−0.0638 (0.7180)	0.0588 (0.8084)	−0.2071 (0.4383)
V	−0.1218 (0.5268)	−0.5029 (0.0514)	0.2417 (0.4228)	0.1377 (0.4220)	0.3438 (0.1691)	0.0598 (0.8083)	0.2407 (0.1783)	0.1487 (0.5398)	0.4036 (0.1310)
Zn	0.3091 (0.1083)	0.3853 (0.1356)	0.3706 (0.2190)	0.0821 (0.6321)	−0.3936 (0.1154)	03.3480 (0.1514)	0.1257 (0.4772)	0.1063 (0.6612)	0.2357 (0.3778)
eBC	**0.4948 (0.0101)**	**0.5353 (0.0118)**	**0.4497 (0.0462)**	0.3224 (0.0610)	0.3456 (0.1669)	0.2611 (0.2817)	0.3105 (0.0790)	0.2074 (0.3924)	0.4679 (0.0800)
UVPM	**0.4226 (0.0281)**	**0.4216 (0.0186)**	0.4308 (0.0524)	0.3076 (0.0729)	0.4578 (0.0669)	0.2340 (0.3341)	**0.3603 (0.0415)**	0.2157 (0.3787)	**0.5357 (0.0450)**

After applying statistics based on matrix correlations assessment, Spearman coefficients and *p*-values (Table 3) at SS it is shown that PHg are correlated with sea salt sources during the whole period (−0.3930, *p*-value = 0.0219 and −0.5532, *p*-value = 0.0376 for Na^+ and Cl^-, respectively) and during the summer season (−0.5196, *p*-value = 0.0377 and −0.4843, *p*-value = 0.0554 for Na^+ and Cl^-, respectively). These correlations could be explained taking in mind the high oceanic influence (backward trajectory analysis in the Supporting Information section) and the reaction of gaseous mercury with the particles of halides carried by marine air mass (see Section 3.2.1). Also, PHg are correlated with crustal and anthropogenic sources; Spearman coefficients and *p*-values between PHg and $SO_4{}^{2-}$ of 0.4476, *p*-value = 0.0091 and 0.6201, *p*-value = 0.0131, for during the whole period and summer season, were achieved. Good correlation between PHg and Cr (an anthropogenic tracer) was also found (0.3467, *p*-value = 0.0432 and 0.4945, *p*-value = 0.0479, for during the whole period and summer season, respectively). On the contrary, the results suggest an enrichment of mercury on PM_{10} due to crustal sources at SS site during winter season (0.5955, *p*-value = 0.0141; 0.5041, *p*-value = 0.0495, 0.5346, *p*-value = 0.0532 and 0.5060, *p*-value = 0.0441 for Ca^{2+}, Al, Fe and Si, respectively). High wind speeds during wintertime would enhance the resuspension of soil/road dust, which could explain the correlation of PHg with crustal components of PM_{10}. Results also suggest an enrichment of mercury on PM_{10} due to anthropogenic/biogenic sources; a good correlation was found with K^+ (0.4778, *p*-value = 0.0488), the presence of K^+ should be attributed to biomass burning in the immediate vicinity of the site.

Finally, crustal and anthropogenic origins for PHg in PM_{10} samples collected at IS could be assigned during the whole period and winter season (Table 3). PHg are positively correlated with Al (0.3457, *p*-value = 0.0505 and 0.6214, *p*-value = 0.0201 for the whole period and winter seasons, respectively), Fe (0.4373, *p*-value = 0.0514 and 0.5964, *p*-value = 0.0380 for the whole period and winter seasons, respectively) and Si (0.3522, *p*-value = 0.0463 and 0.3214, *p*-value = 0.0201 for the whole period and winter seasons, respectively). PHg are also correlated with anthropogenic tracer such as Bi (0.3538, *p*-value = 0.0454 and 0.5332, *p*-value = 0.0460 for the whole period and winter seasons, respectively), Sb (0.4405, *p*-value = 0.0127 and 0.6500, *p*-value = 0.0150 for the whole period and winter seasons, respectively) and UVPM (0.3603, *p*-value = 0.0415 and 0.5357, *p*-value = 0.0450 for the whole period and winter seasons, respectively). No correlation (*p*-value lower than 0.05) were found during summer season at this site.

3.2.2. Principal Component Analysis

PCA has been attempted with a data set in which Na^+, $NH_4{}^+$, K^+, $SO_4{}^{2-}$, $NO_3{}^-$, Al, PHg, Pb, Sb, eBC and UVPM contents were the discriminating variables. The results (after half-range and central value transformation, cross-validation and Varimax rotation) show that 69.71% of the total variance was explained by three principal components (PCs) at US during the whole period, which show eigenvalues higher than 1.0 (Figure 3a). $NH_4{}^+$, $SO_4{}^{2-}$, $NO_3{}^-$ and Pb content (38.67% of the total variance of the data set) are the main features in PC1. PHg, Pb, Sb, eBC and UVPM contents are the main features in PC2, explaining 16.76% of total variance. These PCs are associated with anthropogenic sources. The third PC offers the high weights for Na^+ (sea as source), K^+ (biomass burning) and Al (crustal source) explaining 14.27% of the total variance of the data set, respectively. Similar results were achieved during summer and winter seasons (Figure 3b,c), suggesting an anthropogenic source (fossil fuel combustion and road traffic) for PHg at US during all seasons.

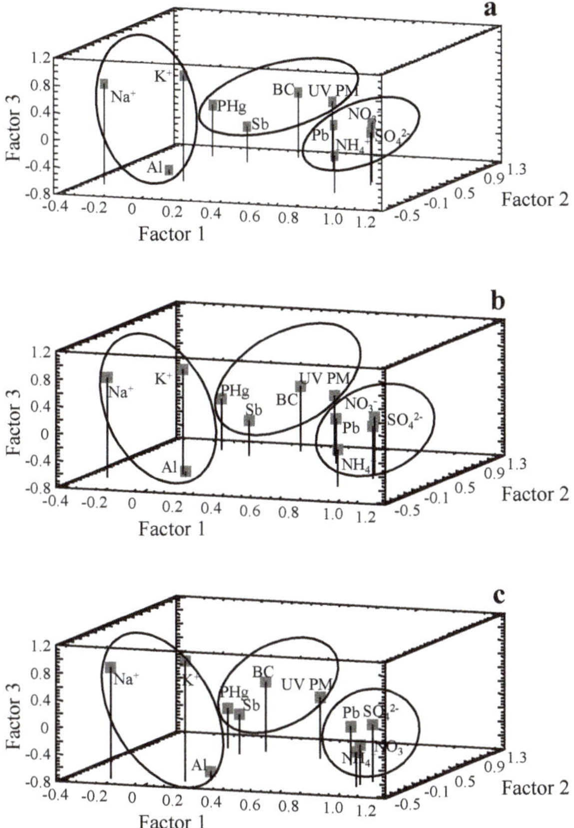

Figure 3. Three-dimensional plots of factor loadings for PM_{10} mass, Na^+, NH_4^+, SO_4^{2-}, NO_3^-, Al, PHg, Pb, Sb, eBC and UVPM contents at urban site during whole period (**a**), summer (**b**) and winter (**c**) seasons.

PCA has also been attempted with PM_{10} samples collected at SS during the whole period and during summer and winter seasons. PCA shown that 72.25%, 81.39% and 78.25% of the total variance was explained by three PCs for the whole period (Figure 4a) and for summer (Figure 4b) and winter (Figure 4c) seasons, respectively. NH_4^+, Al, Sb, eBC and UVPM contents are the main features in PC1 (crustal plus anthropogenic sources), explaining 46.30, 58.67 and 44.86% of total variance at SS during the whole period, and summer and winter seasons, respectively. SO_4^{2-}, NO_3^- and Pb contents (13.77, 14.96 and 18.80% of the total variance of the data set for the whole period, and summer and winter seasons, respectively) are the main features in PC2 (anthropogenic sources). Results shown high loading weights for PHg and K^+ at PC3 for the whole period and winter season (12.17 and 14.58% of the total variance of the data set for the whole period and winter season, respectively), which suggests a biomass combustion source for PHg at SS during the period sampled [35]. Finally, similar factor loadings were achieved for K^+ in the PC1, PC2 and PC3 (0.4960, 0.5567 and 0.4642 for PC1, PC2 and PC3, respectively) at SS during summer season, which can also suggest a biomass combustion source for PHg during this season.

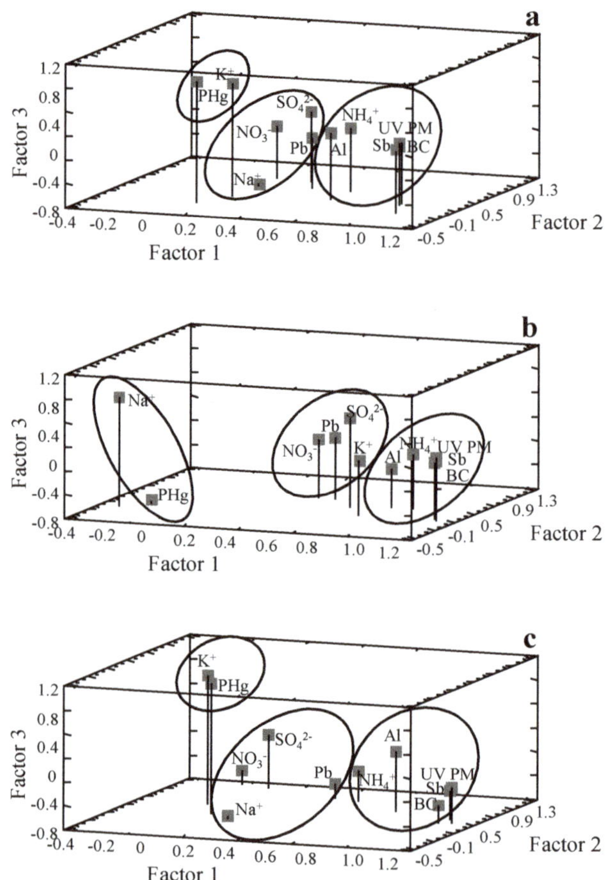

Figure 4. Three-dimensional plots of factor loadings for PM_{10} mass, Na^+, NH_4^+, SO_4^{2-}, NO_3^-, Al, PHg, Pb, Sb, BC and UV PM contents at suburban site during one year long (**a**), summer (**b**) and winter (**c**) seasons.

Finally, PCA has also been attempted with PM_{10} samples collected at IS. Results suggest an anthropogenic origin of PHg during the whole period, a sea salt origin during summer season and anthropogenic and crustal sources during winter season (Figure 5a–c). The marine source of PHg could be explained taking into account that in the coastal atmosphere, Hg^0 is transformed into reactive gaseous mercury ($HgCl_2 + HgBr_2 + HgOBr + \ldots$), which could react or link with the particles of sodium chloride carried by marine air mass [15,36,37]. Backward trajectory analysis shows that the major air masses transported at studied sites during this season come mainly from Atlantic Ocean. This fact have been reported for several anthropogenic compounds by several authors due to the great influence of oceanic air masses in northwest Atlantic European sites [33,38]. As commented in the previous section (Section 3.2.1), the re-suspension of soil/road dust due to high wind speed could explain PHg relationship with crustal components of PM_{10} during winter season.

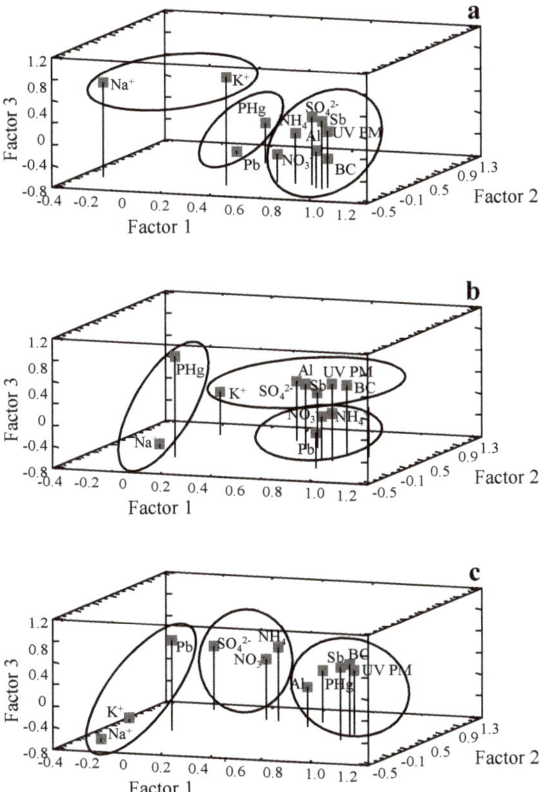

Figure 5. Three-dimensional plots of factor loadings for PM_{10} mass, Na^+, NH_4^+, SO_4^{2-}, NO_3^-, Al, PHg, Pb, Sb, eBC and UVPM contents at industrial site during the whole period (**a**), summer (**b**) and winter (**c**) seasons.

3.3. Human Health Risk Assessment

Non-carcinogenic risk assessment with regard to PHg in PM_{10} was carried out to evaluate the chronic risk of adults by using inhalation chronic daily intake (CDI_{inh}) and hazard quotient (HQ) indexes. The equations used to calculate health risk were based on models recommended by USEPA [39]. CDI_{inh} was assessed by the following equation:

$$CDI_{inh} = \frac{[PHg] \times R_{inh} \times F_{exp} \times T_{exp}}{PEF \times ABW \times T_{avrg}} \quad (1)$$

where CDI_{inh} is the chronic daily intake (mg kg^{-1} day^{-1}), [PHg] is the concentration of particulate mercury in PM_{10} (mg Kg^{-1}), R_{inh} is the inhalation rate (20 m^3 day^{-1} for adults), F_{exp} is exposure frequency (365 day $year^{-1}$), T_{exp} is the exposure duration (24 years for adults), PEF is the particle emission factor (1.36×10^9 m^3 kg^{-1}), ABW is the average body weight (60 kg for adults) and T_{avrg} is the averaging time (for non-carcinogens $T_{avrg} = T_{exp}$). HQ was assessed by the following equation:

$$HQ = \frac{CDI_{inh}}{R_{fD}} \quad (2)$$

where R_{fD} is the reference dose of daily exposure to mercury that is likely to be without deleterious effects (3×10^{-4} mg kg^{-1} day^{-1}) [40]. When HQ ≤ 1 suggests unlikely non carcinogenic effects, HQ > 1 suggests possible non-carcinogenic effects and HQ > 10 indicates high chronic health risk [36]

The mean CDI_{inh} values were $1.5 \times 10^{-5} \pm 1.2 \times 10^{-5}$, $5.7 \times 10^{-5} \pm 7.2 \times 10^{-5}$ and $2.2 \times 10^{-5} \pm 1.4 \times 10^{-5}$ mg Kg^{-1} d^{-1} at US, SS and IS, respectively. The HQ values assessed in PM_{10} at several sites (0.051, 0.189 and 0.073 for US, SS and IS, respectively, Figure 6 were lower than the safe level (HQ = 1), suggesting no non-carcinogenic adverse effects to adults via inhalation.

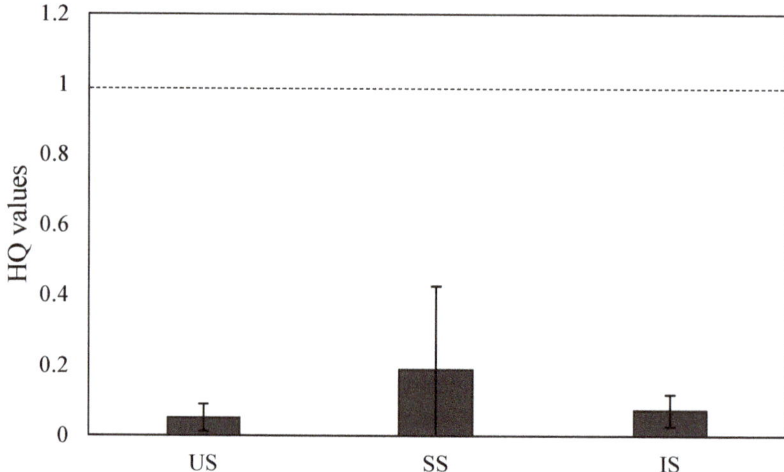

Figure 6. Average Hazard Quotient (HQ) of PHg at urban (US), suburban (SS) and industrial (IS) sites. The dotted line shows the acceptable limit of HQ i.e., 1 as established by US EPA.

4. Conclusions

Our data on PHg levels in PM_{10} at several sites of an Atlantic coastal European region (northwest Spain), represent a novel contribution to the knowledge of complex atmospheric cycle of mercury in regions with high marine influence. In general, it was found that the concentrations of PHg found at these sites were lower than most of other ones reported at continental, Mediterranean and Atlantic coast European sites. PHg in PM_{10} increased, especially due to the burning of fossil fuels for heating purposes. Data from backward trajectory analysis, and univariate and principal component analysis suggest an anthropogenic origin of PHg at urban site during both seasons (summer and winter season). However, at suburban site the main contribution of PHg could be attribute biomass burning. A sea salt and crustal/anthropogenic origin of PHg could be suggested at industrial site. The reaction of Hg^0 emitted with the particles of sodium chloride carried by marine air mass could explain the PHg association with sea salt sources, while the re-suspension of soil/road dust due to high wind speed could explain PHg relationship with crustal components of PM_{10} during winter season. This conclusion could be extrapolated to other north Atlantic urban, suburban and industrial sites of Europe in which the main air masses come from Atlantic Ocean. Finally, the HQ values assessed suggesting no non-carcinogenic risks for PHg at several sites studied.

Supplementary Materials: The following are available online at http://www.mdpi.com/2073-4433/11/1/33/s1, Table S1: Average, RSD, range and minimum and maximum values of major ions, metals, eBC and UVPM in PM_{10} samples at urban site (US) N = 44, suburban site (SS) N = 38 and industrial site (IS) N = 41, Table S2: Seasonal variation of major ions, metals, eBC and UVPM in PM_{10} samples at urban site US (summer season N = 22, winter season N = 22), suburban site SS (summer season N = 17, winter season N = 21) and industrial site IS (summer season N = 18, winter season N = 23).

Author Contributions: M.P.-I., and J.M.-P.; methodology, A.R.-C., M.F.-A., and J.M.-P.; software, J.M.-P.; validation, A.R.-C., M.F.-A., and M.P.-I.; formal analysis, S.M.-L., and J.M.-P.; investigation, A.R.-C., and J.M.-P.; resources,

P.L.-M., and S.M.-L.; data curation, A.R.-C., and J.M.-P.; writing—Original draft, J.M.-P.; writing—Review and editing, J.M.-P., M.P.-I., and M.F.-A.; visualization, J.M.-P., M.P.-I., and A.R.-C.; supervision, P.L.-M., and S.M.-L.; project administration, P.L.-M., and S.M.-L.; funding acquisition, P.L.-M., and S.M.-L. All authors have read and agreed to the published version of the manuscript.

Funding: This work was supported by Xunta de Galicia (Programa de Consolidación y Estructuración de Unidades de Investigación Competitivas ref: ED431C 2017/28-2017-2020) FEDER-MINECO (UNLC15-DE-3097, financed together (80/20%) with Xunta de Galicia and Ministerio de Ciencia, Innovación y Universidades (Programa Estatal de I+D+i Orientada a los Retos de la Sociedad ref: RTI2018-101116-B-I00 (MCIU/AEI/FEDER, UE).

Acknowledgments: We are grateful to Alicia Cantarero-Roldán (SAI-University of A Coruña) for ICP-MS technical support. M. Fernández-Amado appreciates the Ministerio de Ciencia, Innovación y Universidades support (PTA2017-13607-I). The authors would like to thank P. Esperón (PTA2013-8375-I) for her support.

Conflicts of Interest: The authors declare no conflict of interest.

References

1. Pope, C.A., III; Burnett, R.T.; Thun, M.J.; Calle, E.E.; Krewski, D.; Ito, K.; Thurston, G.D. Lung cancer, cardiopulmonary mortality, and long-term exposure to fine particulate air pollution. *J. Am. Med. Assoc.* **2002**, *287*, 1132–1141. [CrossRef]
2. WHO Expert Consultation: Available Evidence for the Future Update of the WHO Global Air Quality Guidelines (AQGs). 2015. Available online: http://www.euro.who.int/__data/assets/pdf_file/0013/301720/Evidence-future-update-AQGs-mtg-report-Bonn-sept-oct-15.pdf?ua=1 (accessed on 15 October 2019).
3. Lindberg, S.E.; Stratton, W.J. Atmospheric mercury speciation: Concentrations and behaviour of reactive gaseous mercury in ambient air. *Environ. Sci. Technol.* **1998**, *32*, 49–57. [CrossRef]
4. EU. Ambient Air Pollution by Mercury (Hg). Position Paper. The European Commission Report. 2001. Available online: http://europa.eu.int/comm/environment/air/background.htm#mercury (accessed on 15 October 2019).
5. The International Air Quality Advisory Board of the International Joint Commission and the Commission for Environmental Cooperation. Addressing Atmospheric Mercury: Science and Policy. 2003. Available online: https://scholar.uwindsor.ca/ijcarchive/558/ (accessed on 15 October 2019).
6. ACAP. *Assessment of Mercury Releases from the Russian Federation*; The Arctic Council Action Plan (ACAP) to Eliminate Pollution of the Arctic; The Ministry of Natural Resources of the Russian Federation, The Danish Environmental Protection Agency and COWI A/S: Copenhagen, Denmark, 2005.
7. ACAP. *Arctic Mercury Releases Assessment. Reduction of atmospheric Mercury Releases from Arctic States*; The Arctic Council Action Plan (ACAP) to eliminate pollution of the Arctic; Danish Environmental Protection Agency and COWI A/S: Copenhagen, Denmark, 2005.
8. United Nations-Environment Programme. Minamata Convention on Mercury. 2017. Available online: http://www.mercuryconvention.org/Portals/11/documents/Booklets/COP1%20version/Minamata-Convention-booklet-eng-full.pdf (accessed on 15 October 2019).
9. Pirrone, N.; Cinnirella, S.; Feng, X.; Finkelman, R.B.; Friedli, H.R.; Leaner, J.; Mason, R.; Mukherjee, A.B.; Stracher, G.B.; Streets, D.G.; et al. Global mercury emissions to the atmosphere from anthropogenic and natural sources. *Atmos. Chem. Phys.* **2010**, *10*, 5951–5964. [CrossRef]
10. Xu, L.; Chen, J.; Yang, L.; Niu, Z.; Tong, L.; Yin, L.; Chen, Y. Characteristics and sources of atmospheric mercury speciation in a coastal city, Xiamen, China. *Chemosphere* **2015**, *119*, 530–539. [CrossRef] [PubMed]
11. Fu, X.; Feng, X.; Qiu, G.; Shang, L.; Zhang, H. Speciated atmospheric mercury and its potential source in Guiyang, China. *Atmos. Environ.* **2011**, *45*, 4205–4212. [CrossRef]
12. Bargagli, R. Atmospheric chemistry of mercury in Antarctica and the role of cryptogams to assess deposition patterns in coastal ice-free areas. *Chemosphere* **2016**, *163*, 202–208. [CrossRef]
13. Han, D.; Zhang, J.; Hu, Z.; Ma, Y.; Duan, Y.; Han, Y.; Chen, X.; Zhou, Y.; Cheng, J.; Wang, W. Particulate mercury in ambient air in Shanghai, China: Size-specific distribution, gase-particle partitioning, and association with carbonaceous composition. *Environ. Pollut.* **2018**, *238*, 543–553. [CrossRef]
14. Tang, Y.; Wang, S.; Wu, Q.; Liu, K.; Wang, L.; Li, S.; Gao, W.; Zhang, L.; Zheng, H.; Li, Z.; et al. Recent decrease trend of atmospheric mercury concentrations in East China: The influence of anthropogenic emissions. *Atmos. Chem. Phys.* **2018**, *18*, 8279–8291. [CrossRef]

15. Xiu, G.L.; Jin, Q.; Zhang, D.; Shi, S.; Huang, X.; Zhang, W.; Bao, L.; Gao, P.; Chen, B. Characterization of size-fractionated particulate mercury in Shanghai ambient air. *Atmos. Environ.* **2005**, *39*, 419–427. [CrossRef]
16. Lewandowska, A.U.; Bełdowska, M.; Witkowska, A.; Falkowska, L.; Wiśniewska, K. Mercury bonds with carbon (OC and EC) in small aerosols (PM1) in the urbanized coastal zone of the Gulf of Gdansk (Southern Baltic). *Ecotoxicol. Environ. Saf.* **2018**, *157*, 350–357. [CrossRef]
17. Pyta, H.; Rogula-Kozłowska, W.; Mathews, B. Co-occurrence of PM2.5-bound mercury and carbon in rural areas affected by coal combustion. *Atmos. Pollut. Res.* **2017**, *8*, 127–135. [CrossRef]
18. Siudek, P.; Frankowski, M.; Siepak, J. Atmospheric particulate mercury at the urban and forest sites in central Poland. *Environ. Sci. Pollut. Res.* **2016**, *23*, 2341–2352. [CrossRef] [PubMed]
19. Beldowska, M.; Saniewska, D.; Falkowska, L.; Lewandowska, A. Mercury in particulate matter over Polish zone of the southern Baltic Sea. *Atmos. Environ.* **2012**, *46*, 397–404. [CrossRef]
20. Li, J.; Sommar, J.; Wängberg, I.; Lindqvist, O.; Wei, S.-Q. Short-time variation of mercury speciation in the urban of Göteborg during GÖTE-2005. *Atmos. Environ.* **2008**, *42*, 8382–8388. [CrossRef]
21. Zielonka, U.; Hlawiczka, S.; Fudala, J.; Wängberg, I.; Munthe, J. Seasonal mercury concentrations measured in rural air in Southern Poland Contribution from local and regional coal combustion. *Atmos. Environ.* **2005**, *39*, 7580–7586. [CrossRef]
22. Wängberg, I.; Munthe, J.; Pirrone, N.; Iverfeldt, Å.; Bahlman, E.; Costa, P.; Ebinghaus, R.; Feng, X.; Ferrara, R.; Gårdfeldt, K.; et al. Atmospheric mercury distribution in Northern Europe and in the Mediterranean region. *Atmos. Environ.* **2001**, *35*, 3019–3025. [CrossRef]
23. Arruti, A.; Fernández-Olmo, I.; Irabien, A. Regional evaluation of particulate matter composition in an Atlantic coastal area (Cantabria region, northern Spain): Spatial variations in different urban and rural environments. *Atmos. Res.* **2011**, *101*, 280–293. [CrossRef]
24. Arruti, A.; Fernández-Olmo, I.; Irabien, A. Evaluation of the contribution of local sources to trace metals levels in urban PM2.5 and PM10 in the Cantabria region (Northern Spain). *J. Environ. Monit.* **2010**, *12*, 1451–1458. [CrossRef]
25. Farinha, M.M.; Almeida, S.M.; Freitas, M.C.; Verburg, T.G.; Wolterbeek, H.T. Influence of meteorological conditions on PM2.5 and PM2.5–10 elemental concentrations on Sado estuary area, Portugal. *J. Radioanal. Nucl. Chem.* **2009**, *282*, 815–819. [CrossRef]
26. Freitas, M.C.; Farinha, M.M.; Ventura, M.G.; Almeida, S.M.; Reis, M.A.; Pacheco, M.G. Gravimetric and chemical features of airborne PM10 and PM2.5 in mainland Portugal. *Environ. Monit. Assess.* **2005**, *109*, 81–95. [CrossRef]
27. Almeida, S.M.; Pio, C.A.; Freitas, M.C.; Reis, M.A.; Trancoso, M.A. Source apportionment of fine and coarse particulate matter in a sub-urban area at the Western European Coast. *Atmos. Environ.* **2005**, *39*, 3127–3138. [CrossRef]
28. Morton-Bermea, O.; Garza-Galindo, R.; Hernández-Álvarez, E.; Ordoñez-Godínez, S.L.; Amador-Muñoz, O.; Beramendi-Orosco, L.; Miranda, J.; Rosas-Pérez, I. Mercury in the metropolitan area of Mexico City. *Bull. Environ. Contam. Toxicol.* **2018**, *100*, 588–592. [CrossRef] [PubMed]
29. Qie, G.; Wang, Y.; Wu, C.; Mao, H.; Zhang, P.; Li, T.; Li, Y.; Talbot, R.; Hou, C.; Yue, T. Distribution and sources of particulate mercury and other trace elements in PM2.5 and PM10 atop Mount Tai, China. *J. Environ. Manag.* **2018**, *215*, 195–205. [CrossRef] [PubMed]
30. Duan, L.; Cheng, N.; Xiu, G.; Wang, F.; Chen, Y. Characteristics and source appointment of atmospheric particulate mercury over East China Sea: Implication on the deposition of atmospheric particulate mercury in marine environment. *Environ. Pollut.* **2017**, *224*, 26–34. [CrossRef]
31. Kumari, A.; Kulshrestha, U. Trace ambient levels of particulate mercury and its sources at a rural site near Delhi. *J. Atmos. Chem.* **2018**, *75*, 335–355. [CrossRef]
32. Xu, H.; Sonke, J.E.; Guinot, B.; Fu, X.; Sun, R.; Lanzanova, A.; Candaudap, F.; Shen, Z.; Cao, J. Seasonal and annual variations in atmospheric Hg and Pb isotopes in Xi'an, China. *Environ. Sci. Technol.* **2017**, *51*, 3759–3766. [CrossRef]
33. Fang, G.-C.; Lo, C.-T.; Zhuang, Y.-J.; Cho, M.-H.; Huang, C.-Y.; Xiao, Y.-F.; Tsai, K.-H. Seasonal variations and sources study by way of back trajectories and ANOVA for ambient air pollutants (particulates and metallic elements) within a mixed area at Longjing, central Taiwan: 1-year observation. *Environ. Geochem. Health* **2017**, *39*, 99–108. [CrossRef]

34. Moreda-Piñeiro, J.; Turnes-Carou, I.; Alonso-Rodríguez, E.; Moscoso-Pérez, C.; Blanco-Heras, G.; López-Mahía, P.; Muniategui-Lorenzo, S.; Prada-Rodríguez, D. The influence of oceanic air masses on concentration of major ions and trace metals in PM2.5 fraction at a coastal European suburban site. *Water Air Soil Pollut.* **2015**, *226*, 2240. [CrossRef]
35. UNE-EN 12341. *Ambient Air-Standard Gravimetric Measurement Method for the Determination of the PM10 or PM2,5 Mass Concentration of Suspended Particulate Matter*; European Committee for Standardization: Brussels, Belgium, 2015.
36. Cheng, I.; Zhang, L.; Blanchard, P.; Dalziel, J.; Tordon, R.; Huang, J.; Holsen, T.M. Comparisons of mercury sources and atmospheric mercury processes between a coastal and inland site. *J. Geophys. Res. Atmos.* **2013**, *118*, 2434–2443. [CrossRef]
37. Hedgecock, I.M.; Pirrone, N. Mercury and photochemistry in the marine boundary layer-modelling studies suggest the in situ production of reactive gas phase mercury. *Atmos. Environ.* **2001**, *35*, 3055–3062. [CrossRef]
38. Sommar, J.; Hallquist, M.; Ljungström, E.; Lindqvist, O. On the gas phase reactions between volatile biogenic mercury species and the nitrate radical. *J. Atmos. Chem.* **1997**, *27*, 233–247. [CrossRef]
39. Santos, P.S.M.; Otero, M.; Santos, E.B.H.; Duarte, A.C. Chemical composition of rainwater at a coastal town on the southwest of Europe: What changes in 20 years? *Sci. Total Environ.* **2011**, *409*, 3548–3553. [CrossRef] [PubMed]
40. US EPA. Risk assessment guidance for superfund. In *Part A: Human Health Evaluation Manual; Part F, Supplemental Guidance for Inhalation Risk Assessment*; U.S. Environmental Protection Agency: Washington, DC, USA, 2011; Volume I.

Article

Household Dust: Loadings and PM_{10}-Bound Plasticizers and Polycyclic Aromatic Hydrocarbons

E. D. Vicente [1], A. Vicente [1], T. Nunes [1], A. Calvo [2], C. del Blanco-Alegre [2], F. Oduber [2], A. Castro [2], R. Fraile [2], F. Amato [3] and C. Alves [1,*]

[1] Centre for Environmental and Marine Studies (CESAM), Department of Environment, University of Aveiro, 3810-193 Aveiro, Portugal; estelaavicente@gmail.com (E.D.V.); anavicente@ua.pt (A.V.); tnunes@ua.pt (T.N.)

[2] Department of Physics, IMARENAB University of León, 24071 León, Spain; aicalg@unileon.es (A.C.); cblaa@unileon.es (C.d.B.-A.); fodup@unileon.es (F.O.); acasi@unileon.es (A.C.); rfral@unileon.es (R.F.)

[3] Institute of Environmental Assessment and Water Research (IDAEA-CSIC), 08034 Barcelona, Spain; fulvio.amato@idaea.csic.es

* Correspondence: celia.alves@ua.pt

Received: 28 October 2019; Accepted: 3 December 2019; Published: 6 December 2019 

Abstract: Residential dust is recognized as a major source of environmental contaminants, including polycyclic aromatic hydrocarbons (PAHs) and plasticizers, such as phthalic acid esters (PAEs). A sampling campaign was carried out to characterize the dust fraction of particulate matter with an aerodynamic diameter smaller than 10 µm (PM_{10}), using an in situ resuspension chamber in three rooms (kitchen, living room, and bedroom) of four Spanish houses. Two samples per room were collected with, at least, a one-week interval. The PM_{10} samples were analyzed for their carbonaceous content by a thermo-optical technique and, after solvent extraction, for 20 PAHs, 8 PAEs and one non-phthalate plasticizer (DEHA) by gas chromatography-mass spectrometry. In general, higher dust loads were observed for parquet flooring as compared with tile. The highest dust loads were obtained for rugs. Total carbon accounted for 9.3 to 51 wt% of the PM_{10} mass. Plasticizer mass fractions varied from 5 $\mu g\ g^{-1}$ to 17 $mg\ g^{-1}$ PM_{10}, whereas lower contributions were registered for PAHs (0.98 to 116 $\mu g\ g^{-1}$). The plasticizer and PAH daily intakes for children and adults via dust ingestion were estimated to be three to four orders of magnitude higher than those via inhalation and dermal contact. The thoracic fraction of household dust was estimated to contribute to an excess of 7.2 to 14 per million people new cancer cases, which exceeds the acceptable risk of one per million.

Keywords: resuspension; household dust; PM_{10}; organic and elemental carbon; phthalic acid esters; polycyclic aromatic hydrocarbons

---

1. Introduction

In industrialized nations, people spend most of their time in closed environments, especially at home [1]. Walking induced particle resuspension has been reported to be an important source of indoor particulate matter [2–4]. Several factors affect resuspension of particles, including relative humidity, flooring type, and dust loadings [3]. Household dust is a complex mixture of particles of both indoor and outdoor origin, including organic, inorganic, and biological components, many of which are toxicants, carcinogens or allergens [5,6]. Its composition depends on numerous conditions, such as environmental and seasonal factors, ventilation and air filtration, homeowner activities, and in- and outdoor sources [7]. Several studies have shown that inhalation of dust particles is linked to an increased risk of a range of health hazards, spanning from asthma symptoms in susceptible adults and children [8–11] to cancer and fertility problems [12].

Residential dust is recognized as a major source of environmental contaminants, including polycyclic aromatic hydrocarbons (PAHs) and phthalic acid esters (PAEs) [13–15]. PAHs are primarily byproducts of incomplete combustion of fossil fuels and biomass and pyrosynthesis of organic materials [16]. Dust, especially in carpeted floor, can be a permanent reservoir for these chemicals, which may be inhaled through resuspension into air, ingested accidentally by children or absorbed through the skin [13,17,18]. Because of their widespread sources and strong carcinogenicity, cytotoxicity, mutagenicity, endocrine disrupting, and other hazardous properties, PAHs have been the focus of extensive attention by scientists and governmental organizations [7,18–24]. Benzo[a]pyrene (BaP), the most extensively studied carcinogenic PAH, is classified by the International Agency for Research on Cancer (IARC) as a Group 1 or known human carcinogen [25]. Four of the top ten priority pollutants, nominated by the Agency for Toxic Substances and Disease Registry (ATSDR) in 2011, are single PAHs or PAH mixtures (PAHs, BaP, benzo[b]fluoranthene, and dibenzo[a,h]anthracene) [26].

PAEs (also called phthalates) are used as plasticizers in several consumer products, commodities, and building materials [27]. Therefore, phthalates are ubiquitous in residential and occupational environments, where they are present in high concentrations, both in air and in dust [14,27–41]. Comparisons of mass fractions reported in different studies indicate that, for measurements conducted over the past decade, the levels of PAEs in indoor dust tend to be three to five orders of magnitude higher than those of PAHs [14]. Recent toxicological studies have proven the potential of PAEs to disturb the human hormonal system and human sexual development and reproduction [42,43]. Moreover, PAEs are suspected to trigger asthma and dermal diseases in children ([44], and references therein). An EU risk assessment classified bis(2-ethylhexyl)phthalate (DEHP), dibutyl phthalate (DBP), and benzyl butyl-phthalate (BBP) as hazardous substances in 2005, and has issued a directive to ban these materials from products, particularly toys and cosmetics [45].

Indoor aerosol sources can significantly contribute to the daily dose of particles deposited into the human respiratory system [46]. Since indoor dust contributes to human exposure, its resuspension rates and chemical composition should be evaluated. The selection of appropriate techniques to assess household dust loadings and composition is a major challenge since several methodologies have been employed, including passive (dust settling) and active techniques (surface wiping, press sampling, sweeping, and vacuuming) [6,47].

Studies on the concentration of contaminants in household dust have been focused on the analysis of the total mass or of sieved fractions, involving sizes of tens or hundreds of μm [17,29,30,32,35,37,39,40]. A summary of average PAH concentrations in settled house dust by country and year (data from 35 studies) can be found in a recent review article [48]. A major issue encountered when comparing these studies was the variability in both the sampling methods employed and the dust particle size fractions subjected to analysis. In 21 out of the 35 studies reviewed, the particle size cut-off points were either 150 μm or 63 μm. It has been suggested that particles >150 μm do not easily and efficiently adhere to hands or skin. Therefore, these sizes are less relevant when evaluating exposure via ingestion or dermal pathways [49]. Liu et al. [24] collected and sieved indoor dust into six size fractions from office and public microenvironments in Nanjing, China. Higher PAH concentrations, and consequently more health risks, were observed for the smallest particles (< 43 μm). In general, the common sampling procedures are affected by the loss of fine particles owing basically to the difficulties of collecting all deposited material and to the electrostatic adhesion of particles to brush hairs, vacuum cleaner bags, and sieve meshes. Considering the possibility that household dust is resuspendable and can become airborne, most methodologies have drawbacks when assessing human inhalation exposure. Such an approach requires measurements of the contaminant concentration in smaller particle sizes. The main goal of this study was to determine household dust loadings and the respective PAH and plasticizer levels in the thoracic fraction (<10 μm) of resuspendable material on the floor, which deposit anywhere within the lung airways. The in situ resuspension chamber was previously devised and successfully applied to collect the deposited particulate matter with an aerodynamic diameter smaller than 10 μm (PM_{10}) from different road pavements [50,51], but it was the first time this active sampling methodology was used to collect settled thoracic particles directly from the floor in indoor environments. For road dust, this sampling technique was previously compared to the USEPA methodology (AP-42 documents), which is based on vacuuming or sweeping of surfaces. Then, the collected material passes through a 200-mesh sieve to determine surface silt loadings, and finally an empirical formula is used to estimate PM_{10} emission factors. Both methodologies provided very comparable results for resuspendable PM_{10} dust [52]. Toxic substances attached to inhaled particles capable of passing beyond the larynx, i.e., PM_{10}, could lead to a series of respiratory and cardiovascular diseases, and increase the risks of cancer [53]. Thus, sampling the total concentration or coarser dust fractions most likely provides only a rough estimate of exposure. In addition to household dust loadings and PM_{10}-bound chemical components, this paper also provides a relatively exhaustive review of literature values to show not only the order of magnitude of the levels, but also the difficulties in comparing results due to the lack of standard methodologies.

2. Methodologies

2.1. Sampling

To determine and characterize dust loadings, a sampling campaign was carried out in four different houses located in the Spanish city of León (Table 1). In each housing unit, three rooms were investigated, including the kitchen, the living room and a bedroom. In each room, two samples were collected with, at least, a one-week interval. Samples in each house were taken one or two days before weekly cleaning. For dust collection, an in situ resuspension chamber operating at an air flow rate of 25 L min^{-1} was used [51]. After vacuuming, PM_{10} was separated from the total dust through a Negretti stainless steel elutriation filter and collected onto 47 mm quartz fiber filters (Pallflex®, Ann Arbor, USA), while particles with aerodynamic diameter >10 μm were deposited in the methacrylate chamber and along the elutriation filter. Electrostatic adhesion could cause some losses of particles <10 μm. However, this loss is likely to be negligible with respect to losses of traditional sampling procedures (i.e., sweeping). In a previous work, the chamber sediments were brought to a laboratory, dried for 48 h at room temperature, sieved and, then, analyzed by means of an optical particle sizer with the aim of verifying the granulometry selection of the sampling system. Results showed that the fraction <10 μm was, on average, only 0.6% and 0.1% (in volume) of samples previously sieved at 250 μm and 63 μm, respectively [50].

Table 1. Characteristics of the houses where floor dust particulate matter with an aerodynamic diameter smaller than 10 μm (PM_{10}) was sampled.

House	Characteristics	Room	Flooring
1	Suburban two-story house with well-ventilated kitchen, two occupants	Kitchen	Tile
		Bedroom	Parquet
		Living room	Parquet
		Living room rug	Cut pile carpet/rug
		Bedroom rug	Long threads shag rug
2	Single story apartment located in the city center, two occupants	Kitchen	Tile
		Bedroom	Parquet
		Living room	Parquet
3	Rural two-story house with open fireplace in the living room, two occupants	Kitchen	Tile
		Bedroom	Parquet
		Living room	Tile
4	Single story apartment with small kitchen open to the living room, one occupant	Kitchen	Tile
		Bedroom	Tile
		Living room	Tile

Sampling was performed in surface areas of 1 m^2 for 30 min. Two to three different square meters were sampled using the same filter in order to ensure enough particulate mass for the subsequent gravimetric and chemical analyzes. Since some compounds could derive from the resuspension chamber itself or from ambient air that enters the system and passes through the filter during dust collection from the floor, air samples were vacuumed through the same system after sampling in each room. These background air filters were also sampled for 30 min.

2.2. Household Dust Characterization

Due to the loss of a small fragment after sampling, the kitchen filter from house 2 obtained in the 2nd week was discarded because weighing led to a negative dust mass. After gravimetric determination, two punches (9 mm) of each filter were analyzed by a thermo-optical transmission technique to obtain the PM_{10} carbonaceous content (organic and elemental carbon, OC and EC). This method is based on the quantification of the CO_2 released from the volatilization and oxidation of different carbon fractions under controlled heating by a non-dispersive infrared (NDIR) analyzer. The blackening of the filter is monitored using a laser beam and a photodetector, which enables separating the EC formed by pyrolysis [54]. The remaining portion of each filter was extracted by sonication for 15 min with three aliquots (25 mL each) of dichloromethane. After filtration, the solvent was concentrated in a TurboVap system from Biotage and evaporated to dryness by a gentle nitrogen stream. All the extracts were analyzed by gas chromatography-mass spectrometry (GC-MS) in a Shimadzu QP5050A equipped with a TRB-5MS 30 m × 0.25 mm × 0.25 μm column. The quantitative analysis was performed by single ion monitoring (SIM). Background air filters were analyzed in the same way as the samples to obtain blank-corrected results. Data were acquired in the electron impact (EI) mode (70 eV). The oven temperature program was as follows: 60 °C (1 min), 60 to 150 °C (10 °C min^{-1}), 150 to 290 °C (5 °C min^{-1}), 290 °C (30 min) and using helium as carrier gas at 1.2 mL min^{-1}. For the quantification of PAHs, the following mixture of deuterated internal standards (IS) was used: 1,4-dichlorobenzene-d4, naphtalene-d8, acenaphthene-d10, phenanthrene-d10, chrysene-d12, and perylene-d12 (Supelco, St. Louis, USA). In the case of plasticizers, deuterated diethyl phthalate-3,4,5,6-d4 and bis(2-ethylhexyl)phthalate-3,4,5,6-d4 (Supelco, St. Louis, USA) were used as IS. Calibrations were performed with authentic standards (Sigma-Aldrich, St. Louis, USA) in eight different concentration levels.

2.3. Extraction Recoveries

To assess recoveries during extraction, prebaked blank filters were spiked with known amounts of standards, covering the concentration range commonly reported in the literature for household dust. The solutions of known concentrations were applied to the entire surface of each filter with Pasteur pipettes. The solvent was allowed to evaporate keeping the filters in a desiccator overnight. The filters were subjected to the same methodology of extraction and analysis used for the samples. Six prebaked blank filters were also extracted and analyzed. Four distinct concentrations were tested, in triplicate, and each extract was injected 3 times. The following overall recoveries (%) were obtained: diethyl phthalate 38.4 ± 8.05, di-n-butyl phthalate 113 ± 17.4, benzyl butyl phthalate 70.2 ± 12.7, bis(2-ethylhexyl) adipate 70.4 ± 12.9, bis(2-ethylhexyl) phthalate 108 ± 10.9, phenanthrene 53.0 ± 10.6, anthracene 67.1 ± 11.5, fluoranthene 84.7 ± 11.2, pyrene 81.9 ± 11.0, benzo[a]anthracene 97.2 ± 11.4, chrysene 81.9 ± 9.7, benzo[b]fluoranthene 106 ± 10.0, benzo[k]fluoranthene 108 ± 10.6, benzo[a]pyrene 93.4 ± 13.3, indeno [1,2,3-cd]pyrene 90.6 ± 15.8, dibenzo[a,h]anthracene 87.2 ± 16.7, and benzo[g,h,i]perylene 84.7 ± 16.5. The concentrations shown throughout the manuscript were not adjusted with these percent recoveries. Due to its high volatility and concentration variability in both dust and background air samples, naphthalene was excluded from quantification. Due to its vapor pressure of 0.087 mm Hg at 25 °C, naphthalene is sometimes considered a "borderline" volatile/semivolatile compound, since it may often be detected in both gas and particulate phases. Because of its tendency to sublimate, in ambient air, naphthalene is known to mainly exist in the vapor phase [55].

2.4. Health Risk Evaluation

Human exposure to plasticizers occurs via ingestion, inhalation, and dermal contact. The daily intake (DI) through different pathways were estimated by the following equations ([39], and references therein), where DI_{ing}, DI_{inh} and, DI_{der} are the daily intakes (ng kg^{-1} day^{-1}) via dust ingestion, inhalation, and dermal contact, respectively:

$$DI_{ing} = \frac{C_{dust} \times IngR \times EF \times ED \times CF}{BW \times AT} \quad (1)$$

$$DI_{inh} = \frac{C_{dust} \times InhR \times EF \times ED}{BW \times AT \times PEF} \quad (2)$$

$$DI_{der} = \frac{C_{dust} \times SA \times AF_{dust} \times ABS \times EF \times ED \times CF}{BW \times AT} \quad (3)$$

C_{dust} represent the mean concentrations in dust (ng g^{-1}); IngR is the ingestion rate of indoor dust (200 mg day^{-1} for children, 100 mg day^{-1} for adults); EF is the exposure frequency (180 days per year for both children and adults); ED is the exposure duration (6 years for children, 24 years for adults); CF is an unit conversion factor (10^{-3} g mg^{-1}); BW is the body weight (15 kg for children, 70 kg for adults); AT is the average time (2190 days for children, 8760 days for adults); InhR is the inhalation rate (7.6 m^3 day^{-1} for children, 12.8 m^3 day^{-1} for adults); PEF is the particulate emission factor (1.36×10^6 m^3 g^{-1}); SA is the dermal exposure area (1150 cm^2 for children, 2145 cm^2 for adults); AF_{dust} is the dust adherence factor (0.2 mg cm^{-2} day^{-1} for children, 0.07 mg cm^{-2} day^{-1} for adults); and ABS is the dermal adsorption fraction (0.001 for both children and adults, dimensionless).

Like plasticizers, PAHs can enter the body through ingestion (swallowing), inhalation (breathing), and skin contact. The carcinogenic potency of PAHs in indoor dust can be evaluated by comparing the carcinogenic potency of each PAH to that of BaP. The BaP carcinogenic equivalent concentration (BaP_{TEQ}) is defined as follows:

$$BaP_{TEQ} = \Sigma C_i \times BaP_{TEF} \quad (4)$$

where C_i is the concentration of each individual PAH and BaP_{TEF} are toxic equivalency factors [24]. The incremental lifetime cancer risk (ILCR) is generally used to quantitatively estimate the exposure risks from the three exposure routes [24,56]:

$$ILCR_{ing} = \frac{CS \times \left(CSF_{ing} \sqrt[3]{BW/70}\right) \times IngR \times EF \times ED \times CF}{BW \times AT} \quad (5)$$

$$ILCR_{inh} = \frac{CS \times \left(CSF_{inh} \times \sqrt[3]{BW/70}\right) InhR \times EF \times ED}{BW \times AT \times PEF} \quad (6)$$

$$ILCR_{der} = \frac{CS \times \left(CSF_{der} \times \sqrt[3]{BW/70}\right) SA \times AF_{dust} \times ABS \times EF \times ED \times CF}{BW \times AT} \quad (7)$$

where CS is the BaP_{TEQ} concentration (mg kg^{-1}); CSF_{ing}, CSF_{inh}, and CSF_{der} are carcinogenic slope factors of 7.3, 3.85, and 25 $(mg\ kg^{-1}\ day^{-1})^{-1}$, respectively.

The risk associated with non-carcinogenic PAHs in household PM_{10} dust was estimated through the hazard quotient (HQ):

$$HQ = DI_{nc}/R_fD \quad (8)$$

where DI_{nc} is the daily intake via ingestion of non-carcinogenic PAHs, obtained by Equation (1), and R_fD is the reference dose. The oral ingestion R_fD values were taken from Iwegbue et al. [57].

3. Results and Discussion

3.1. Dust Loadings

Huge differences in dust loadings between rugs and hard floorings were registered (Figure 1). Rugs represented the surface with the highest dust loadings. In general, higher dust loadings were observed for parquet flooring as compared with tile. Among the four dwellings, the highest dust loadings were obtained in the living room of a suburban family house with an open fireplace. Several factors can affect the surface dust loadings, such as walking, cleaning frequency, household materials, and indoor particle sources. Thus, the presence or absence of occupants can have a marked effect on resuspension levels. In house 1, for example, a very significant decrease in dust loadings from the first to the second week was registered, possibly due to the absence of owners for a few days. The highest mean value was observed in bedrooms (1745 μg PM_{10} m^{-2}), followed by living rooms (785 μg PM_{10} m^{-2}) and, finally, kitchens (297 μg PM_{10} m^{-2}). The size distribution of aerosols emitted from cooking activities has been reported to be dominated by ultrafine particles, with modes generally in the range of 20–100 nm [58]. Thus, although cooking emissions can be high and contribute to the deposition and subsequent resuspension of dust, nanometer sizes represent a small fraction of the particulate mass. Dust loadings in the kitchen of house 3 were two to seven times higher than in the kitchens of other houses. This may be due to a higher utilization rate. In the other three residences the owners, in general, only prepare dinner. A higher mean value for bedrooms may be related to the presence of specific active sources or activities in this space of the house, such as making and unmaking the bed, dressing and undressing (which potentiates skin desquamation), existence of various rugs, use of cosmetics, etc. [59].

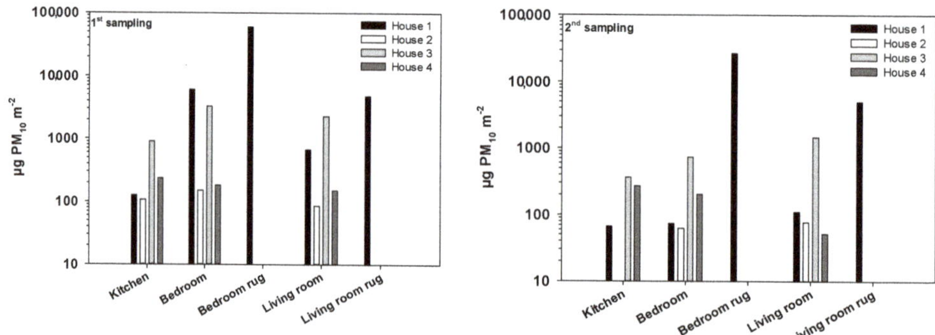

Figure 1. Dust loadings in the different rooms of the various houses for the two sampling campaigns.

Using a mechanical resuspension device, Tian et al. [3] characterized walking-induced particle resuspension as a function of flooring type. Results showed that for particles at 0.4 to 3.0 μm, the difference in resuspension fraction between carpets and hard floorings was not significant. It was also found that for fine particles (0.4 to 3.0 μm), the difference in resuspension caused by flooring type is negligible, while for coarse particles (3.0 to 10 μm) carpets are associated with two to four times higher resuspended concentration in comparison with hard floorings. In fact, carpeted floors may contribute to significantly higher surface dust loadings and allergen concentrations than hard floors [60–64]. Roberts et al. [65] used a high-volume surface sampler to measure surface dust in carpets. Dust loadings ranged from 0.32 to 14.4 g m^{-2}. Adgate et al. [66] collected bare floor and carpet dust samples in 216 Jersey City, New Jersey, homes using quantitative wipe and vacuum sampling techniques. Dust loadings varied from 0.05 and 7.0 g m^{-2} and from 0.3 and 99 g m^{-2} in the wipe floor and vacuum samples, respectively. It should be noted that dissimilar indoor dust sampling strategies (e.g., wipe versus vacuum methods) are used to measure loadings and the amounts of toxicants per unit area, which renders comparisons between studies difficult. Lioy et al. [67] found that while loadings were substantially greater with wipe sampling, metal concentrations within the dust samples were similar for both methods of sampling. Vacuum cleaner sampling has its own series of problems, especially the variability in design and efficiency, and likely does not retain particles below 10 μm [47]. Bai et al. [68] evaluated the following five methods of sampling lead-contaminated dust on carpets: (i) wipe, (ii) adhesive label, (iii) C18 sheet, (iv) vacuum, and (v) hand rinse. The wipe and vacuum methods showed the best reproducibility and correlation with other sampling techniques. The authors concluded that surface wipe sampling was the best method to measure accessible lead from carpets for exposure assessment, while vacuum sampling was most effective for providing information on total lead accumulation (long-term concentrations). In their review paper, Lioy et al. [6] stated, "although we have come a long way in determining the uses of house dust to identify sources of indoor contamination and to provide improved estimates of residential human exposure, one of the challenges still lies in the reliability of sampling techniques." The instrument used in this study has the advantage of being a device capable of collecting dust particles below 10 μm, for gravimetric and chemical analysis. Sampling and analysis of higher dust particle sizes may only provide a crude estimate of inhalation exposure. However, validation and intercomparison studies are still needed, which is not easy, as the panoply of established methodologies does not match inhalable sizes.

3.2. Carbonaceous, Plasticizer, and PAH Particulate Mass Fractions

Total carbon accounted for 9.3 to 51 wt% of PM_{10} with the highest mass fractions recorded in dust samples collected in the city center apartment (Figure 2). More than 80% of the total carbonaceous matter was composed of OC, whereas in many samples the EC was too low or undetectable. The highest percentages of EC (10% to 20% of total carbon) were observed in kitchens, where there are sources of combustion. The high proportion of OC reflects the importance of a multitude of indoor sources that contribute to the organic carbonaceous component of household dust, including bacteria, skin flakes, cosmetics, cleaning products, cooking, paper and clothing fibers, microscopic specks of plastics, environmental contaminants brought on the soles of our shoes, etc. Our finding that a large portion of PM_{10} from indoor dust is composed of OC agrees with the results of Polidori et al. [69], who measured particulate OC and EC concentrations at 173 homes in the USA. These authors demonstrated that part of the OC can be secondarily formed in the indoor environment as a result of reactions involving gas-phase organic compounds emitted by cleaning products, air fresheners, and other sources.

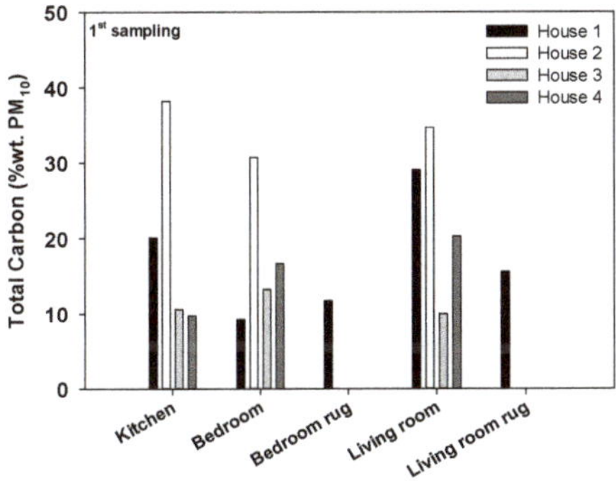

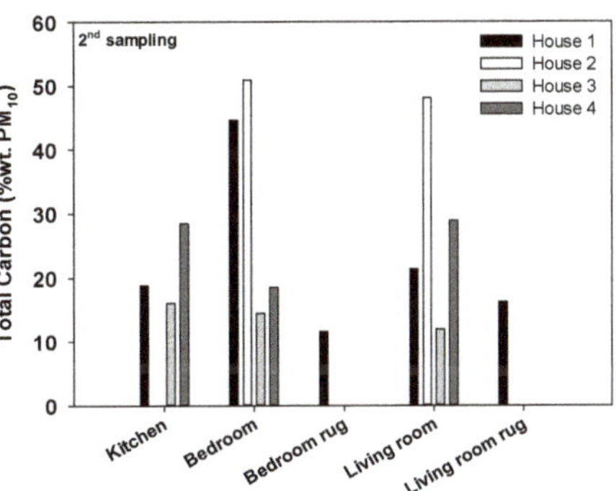

Figure 2. Total carbon (TC = OC + EC) mass fraction in resuspendable PM_{10} from household dust.

Eight phthalate plasticizers (PAEs) and one non-phthalate plasticizer [bis(2-ethylhexyl) adipate, DEHA] were quantified in PM_{10} from settled house dust. PAEs included dimethyl phthalate (DMP), dimethylpropyl phthalate (DMPP), diethyl phthalate (DEP), diisobutyl phthalate (DIBP), di-n-butyl phthalate (DBP), di-n-hexylphthalate (DNHP), benzyl butyl phthalate (BBP), dicyclohexyl phthalate

(DCHP), bis(2-ethylhexyl) phthalate (DEHP), di-n-octyl phthalate (DNOP), di-isononylphthalate (DINP), and di-isodecylphthalate (DIDP). The total mass fractions varied from 5 μg g^{-1} to 17 mg g^{-1} PM_{10} (Figure 3). As observed with dust loads, huge differences were observed from week to week and from home to home. The highest values were registered in the bedroom of the city center apartment and in the living room of the rural house. In addition to variability in sources, the concentrations of these compounds depend on the same factors already mentioned for the dust loads (e.g., ventilation, domestic routines, and cleaning activities). Widely scattered concentration levels were also documented in many previous works. For example, Kubwabo et al. [33] analyzed 17 PAEs in 126 Canadian household dust samples, evidencing the huge variability in spatial and temporal distribution of these compounds across different areas of the home, and thus the difficulty in predicting potential household exposures.

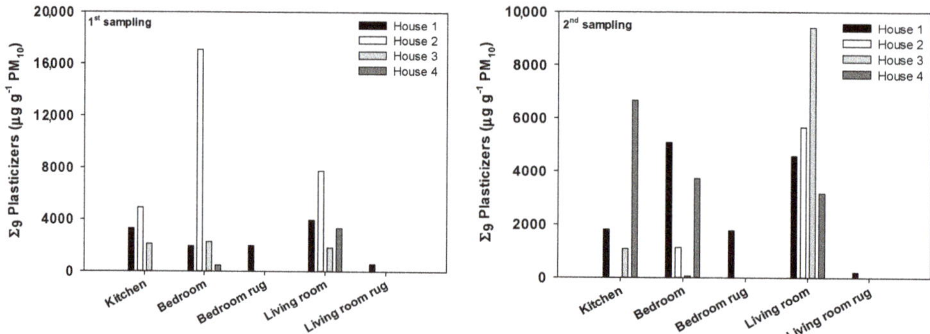

Figure 3. Plasticizer mass fractions in PM_{10} from resuspended household dust.

Most likely due to its high volatility, DMP was the compound with the lowest mass fractions (Table 2). On the other hand, DEHP, DNOP, and DBP were the major PAEs in household dust. While high DBP values were observed in all parquet floor bedrooms, only samples from two living rooms showed detectable masses. Bamai et al. [70] also associated higher DBP levels in floor dust with compressed wooden floor. This type of flooring is usually composed of thin pieces, which are glued together and covered with wax, paint, and sometimes flame retardants. The surface applied products (gloss agents, plastic additives, paint, and varnish) contain DBP [70]. From quantitative and qualitative emission data on phthalates from different materials, Afshari et al. [71] reported that polyolefin covered with wax for floor polishing increased DBP concentration in chamber air by two-fold. DBP is also employed as a coalescing aid in latex adhesives, as well as a plasticizer in cellulose plastics and a solvent for dyes [70]. Furthermore, DBP has been reported to be largely present in cosmetic and personal care products [72].

Although the concentration of DNOP in floor dust has been reported in only a very limited number of publications, the mass fractions of this study are higher than those described in the literature. There were no appreciable differences between the amounts found in PM_{10} of the various rooms of the houses. DNOP is used in carpet back coating, floor tile, and adhesives. It is also employed in cosmetics and pesticides [73]. DEHP was present at higher concentrations in the bedroom and living room samples. Although DEHP has been consistently described as one of the most abundant PAEs in settled dust, the levels have decreased over time, reflecting its phase-out in the EU [28]. In Europe, the use of DEHP decreased drastically in 2001 and has to a large extent been replaced by DINP and DIDP, with their longer chains and lower volatility [74]. DEHP has been used in numerous consumer products, children toys, medical devices, and building materials (e.g., vinyl flooring, furniture, paints, cables, wires, wall coverings, and packaging materials) [75]. Bamai et al. [70] and Bornehag et al. [34] associated polyvinyl chloride (PVC) flooring with DEHP levels in house dust. Kolarik et al. [76] documented higher concentrations of BBP, DNOP, and DEHP in indoor dust in homes where polishing agents were employed as compared with homes where such products were infrequently used or not used at all. As PVC flooring was not present in any of the houses for this study, other emission sources may have contributed to the DEHP levels. It has been shown that the source characteristics (surface area and material phase concentration of DEHP), as well as the external mass-transfer coefficient and ventilation rate, are important variables that influence the steady-state DEHP concentration and the resulting exposure [77]. DEHP and other PAEs are strongly sorbed to surfaces. A relatively small gas-phase concentration, such as 0.1 ppb, is enough for significant vapor transport of a PAE and its subsequent partitioning between the gas phase and indoor surfaces, including airborne particles and settled dust [78].

BBP was present in all samples at relatively similar concentrations regardless of room type. The values fall within the range reported for household dust from other regions. BBP is commonly employed as a plasticizer for vinyl foams, which are often used as floor tiles. Other uses are in artificial leather, paints, and adhesives. BBP was classified as toxic by the European Chemical Bureau, and therefore its use has decayed rapidly in the last decade [79].

No appreciable differences were found between DINP concentrations in samples from the various rooms of the houses. DIPD, instead, was present in higher amounts in living rooms and at lower levels in kitchens. The values observed in this study for these two compounds seem to be lower than those documented by Santillo et al. [80] for dust samples of houses in several European cities, although the ranges reported by these researchers are very wide. On the basis of a risk assessment, in 2013, the European Chemicals Agency (ECHA) concluded that there is no evidence that would justify a re-examination of the existing restriction on DINP and DIDP in toys and childcare articles which can be placed in the mouth by children [81]. About 95% of DINP is used in PVC applications. The other 5% is employed in non-PVC applications such as rubbers, adhesives, sealants, paints and lacquers, and lubricants. For DIDP, non-PVC applications are comparatively small, but comprise use in anticorrosion and antifouling paints, sealing compounds and textile inks ([81], and references therein).

Table 2. Comparison of mass fractions of plasticizers (µg g^{-1}) in house dust.

Sampling Information	Statistics	DMP	DMPP	DEP	DIBP	DBP	DNHP	BBP	DEHA	DCHP	DEHP	DNOP	DINP	DIDP	Ref.
Kitchens	Mean	0.27		142		562		6.3	17		271	927	10	6.6	This study
	min-max	<0.1–1.89		<0.1–567		<0.1–3301		<0.1–12	<0.1–97		<0.1–795	<0.1–3270	<0.1–25	<0.1–46	
	Median	<0.1		<0.1		<0.1		6.4	<0.1		<0.1	407	9.0	<0.1	
Bedrooms	Mean	0.33		45		2353		10.4	36		897	969	20	35	This study
	min-max	<0.1–1.57		<0.1–354		<0.1–9397		4.25–26	<0.1–191		<0.1–2864	<0.1–2594	4.15–78	<0.1–215	
	Median	0.19		<0.1		964		7.0	7.4		442	316	7.3	3.9	
Living rooms	Mean	0.93		139		1181		11	19		1352	751	28	66	This study
	min-max	<0.1–4.5		<0.1–924		<0.1–6044		5.1–21	<0.1–109		<0.1–8374	<0.1–2468	<0.1–145	<0.1–300	
	Median	<0.1		16		<0.1		9.3	<0.1		218	473	8.2	16.6	
Living room rug *	Mean	0.60		21		893		8.54	23.4		1997	807	9.8	23.8	This study
Bedroom rug *	Mean	1.31		5.8		139		3.44	3.7		92	119	5.3	9.1	This study
Dust samples (<180 µm) from houses of China collected with brushes	Mean	0.693		0.187	17.1	26.4	0.003	0.040		0.015	105	0.342			[39]
	Median	0.181		0.116	9.33	12.9	0.001	0.014		0.007	25.6	0.130			
30 household dust samples (<63 µm) from vacuum cleaner	Median					87.4		15.2			604		129	33.6	[32]
Dust from 30 apartments in bags of vacuum cleaner	Mean	10.8	54.6	44.6		55.6		86.1			776				[82]
	Median	1.5	37.5	6.1		47.0		29.7			703				
Dust (<250 µm) from 11 houses, 3 labs and 1 hospital in vacuum cleaner bags, Qatar	Mean	0.89		7.9		12.5	1.6	2.6	10.7		288	5.8	106	11.4	[35]
	Median	0.98		7.0		17.0	0.12	1.95	9.6		395	3.0	101	11.0	
Dust samples (<250 µm) from homes in Kuwait from vacuum cleaner bags	Mean	0.01		1.5		51	0.2	6.4		0.3	1700	14			[83]
	Median	0.03		1.8		45	0.39	8.6		2.9	2256	14			
House dust from China collected by sweeping the floor and wiping the top of furniture (<2 mm)	Median	0.2		0.4	17.2	20.1	nd	0.2		nd	228	0.2			[40]
Dust from Albany, USA, from vacuum cleaner bags of several homes (<2 mm)	Median	0.08		2.0	3.8	13.1	0.6	21.1		nd	304	0.4			[40]
Saudi floor dust from vacuum cleaner bags (<250 µm)	Mean	1.4		4.2	33.6	80.2		1.5			1140	102.4			[37]
	Median	0.6		1.4	22.1	33.3		0.8			1020	26.8			
Kuwaiti floor dust from vacuum cleaner bags (<250 µm)	Mean	0.2		5.1	20.0	4.0		1.3			220	2.4			[37]
	Median	0.1		2.7	17.2	1.6		0.8			240	2.8			

Table 2. *Cont.*

Sampling Information	Statistics	DMP	DMPP	DEP	DIBP	DBP	DNHP	BBP	DEHA	DCHP	DEHP	DNOP	DINP	DIDP	Ref.
Settled dust collected in child's room, above the floor level, in Bulgarian homes by vacuum cleaner	Mean	260		350		7850		320			960	250			[76]
	Median	280		340		9930		340			1050	300			
Dust from rooms in Germany collected with vacuum cleaner (<2 mm)	Mean	1.32		80.7	107.1	384		85.8			1026		114	13.5	[80]
	Median	1.42		12.9	36.5	44.1		82.2			996		113	< 0.1	
Dust from rooms in Spain (Madrid) collected with vacuum cleaner (<2 mm)	Mean	0.14		14.0	265	131		6.55			464		69.5	10.2	[80]
	Median	< 0.1		7.88	201	145		5.3			370		<0.1	<0.1	
Dust from rooms in France (Paris) collected with vacuum cleaner (<2 mm)	Mean	<0.1		9.88	91.4	174		227			111		122	48.2	[80]
	Median	<0.1		8.72	86.1	65.8		200			356		115	<0.1	
Dust from rooms in Italy (Rome) collected with vacuum cleaner (<2 mm)	Mean	0.30		9.60	221	36.6		89.0			503		76.0	142	[80]
	Median	<0.1		6.78	180	42.8		23.6			434		<0.1	<0.1	
Dust from rooms in the UK collected with vacuum cleaner (<2 mm)	Mean	0.12		12.2	52	50.2		56.5			192		48.5	20.8	[80]
	Median	<0.1		3.5	43.2	52.8		24.5			195		<0.1	<0.1	
Settled dust from apartments in Stockholm from vacuum cleaner	Median	0.47		14	104	103		16			449	0.00	106	56	[74]
Dust from the floor surface and from objects within 35 cm above the floor with vacuum cleaner in Japanese dwellings	Median	<0.5		< 0.5	3.1	16.6		2.0	8.0		1110		139		[70]
Floor dust samples collected with vacuum cleaner from living rooms in 41 dwellings, Sapporo (Japan)	Median	<0.2		0.33	2.9	19.8		4.2	6.5		880		126		[84]
Dust samples from household vacuum cleaner bags in French dwellings (<100 μm)	Mean	0.26		10.6	111	10.5		25.8			441		158		[17]
	Median	0.25		4.9	20	9.1		6.1			462		139		
House dust from urban dwellings in Nanjing, China, collected with vacuum cleaner (150 μm)	Mean	0.4		0.9		52.3		2.9			462	1.6			[36]
	Median	0.1		0.2		23.7		1.6			183	0.1			
Dust from homes across the USA with vacuum cleaner (<1.4 mm)	Mean	0.05		2.34	6.24	4.17		214	77.1		97.2	144			[29]
	Median	0.04		0.49	4.08	3.54		51.2	21.8		73.1	74.9			

* only two samples, dimethyl phthalate (DMP), dimethylpropyl phthalate (DMPP), diethyl phthalate (DEP), diisobutyl phthalate (DIBP), di-n-butyl phthalate (DBP), di-n-hexylphthalate (DNHP), benzyl butyl phthalate (BBP), bis(2-ethylhexyl) adipate (DEHA), dicyclohexyl phthalate (DCHP), bis(2-ethylhexyl) phthalate (DEHP), di-n-octyl phthalate (DNOP), di-isononylphthalate (DINP), and di-isodecylphthalate (DIDP); nd, not detected.

As compared with plasticizer compounds, PAHs accounted for a much smaller mass of household dust (Σ_{20}PAHs 0.98 to 116 µg g^{-1} PM$_{10}$). Given the small number of samples and the variability in concentrations, it is difficult to infer patterns between house or room typologies (Figure 4). The values obtained in this study fall into the broad range of values reported in the literature. Wang et al. [22] analyzed 15 PAHs in settled house dust of urban dwellings with preschool-aged children in Nanjing, China. $\sum_{15}$PAHs ranged from 1.2 to 280.4 µg g^{-1}, averaging 11.1 µg g^{-1}. Yadav et al. [85] investigated the contamination level of EPA's priority PAHs in indoor dust from residential, educational, commercial, public places, and office premises, in four major cities of Nepal. Concentrations of $\sum_{16}$PAHs ranged from 747 to 4910 ng g^{-1} (median 1320 ng g^{-1}). In Palermo, Italy, Mannino and Orecchio [86] collected indoor dust samples by brushing from surfaces at a height of 1.5 to 2.0 m above ground level in bedrooms, living rooms, kitchens, laboratories, offices, in a market, and in a car. The $\sum_{16}$PAH concentrations were within a broad interval (36 to 34,453 µg g^{-1}), with an average of 5111 µg g^{-1}, indicating heterogeneous levels of contamination in the investigated microenvironments. Organic extracts of sieved vacuum cleaner dust from 51 homes in Canada were examined for the presence of 13 PAHs [87]. Total concentrations varied between 1.5 and 325 µg g^{-1} with a geometric mean of 12.9 µg g^{-1}. These values were found to be comparable to those documented in a previous review in which the total PAHs in samples collected from urban, rural, and suburban homes ranged between 0.4–544 µg g^{-1} with a geometric mean of 4.5 µg g^{-1} [7]. High concentrations of $\sum_{16}$PAHs were also observed in indoor dust samples collected across China from 45 private domiciles and 36 public buildings (1.00 to 470 µg g^{-1}, mean value of 30.9 µg g^{-1}) [88]. It must be noted, once again, that comparability between results of various works should be made with caution, as they concern different surfaces, particle sizes, sampling and analytical methodologies, and list of compounds.

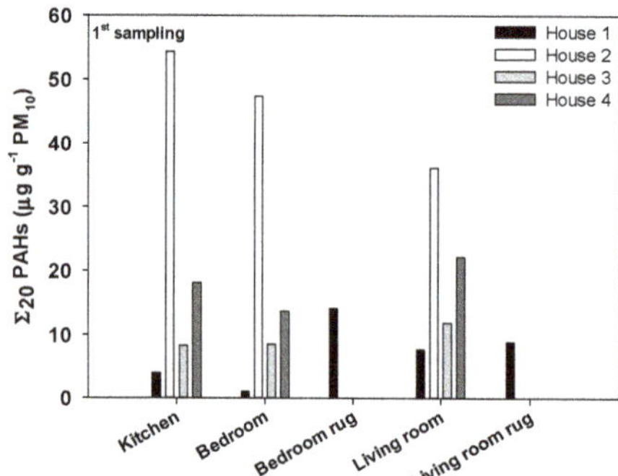

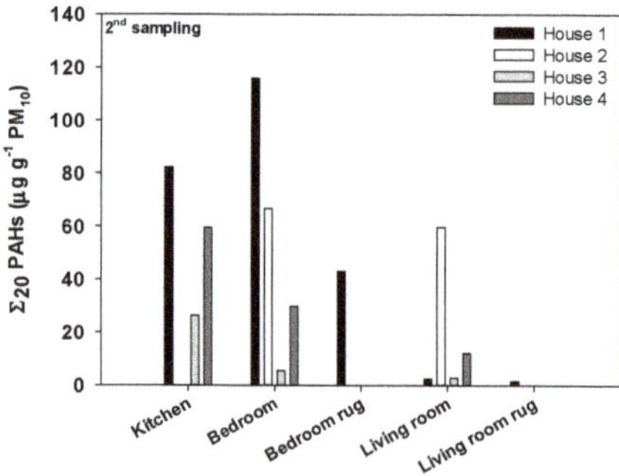

Figure 4. Polycyclic aromatic hydrocarbon (PAH) mass fractions in PM$_{10}$ dust samples collected in two different sampling periods.

Low molecular weight-PAHs (LMW, two and three rings) were less abundant than high molecular weight-PAHs (HMW, four and six rings), suggesting the dominance of pyrogenic sources. Regardless of the microenvironment, the median LMW/HMW ratios were always in the range from 0.3 to 0.5. The main PAHs in the thoracic fraction of resuspended dust were pyrene, retene, and indeno[1,2,3-cd]pyrene, reaching concentrations up to 27, 21, and 13 µg g^{-1}, respectively (Table 3). While the medians of the latter two compounds were higher in the PM$_{10}$ sampled in the living rooms, pyrene showed higher levels in the kitchens. PAH levels and speciation are highly dependent on the cooking methods [58], biomass burning appliances and operating conditions [89], traffic fleet and meteorology in the outdoor surrounding environment [90], among other factors.

Table 3. Comparison of PAH levels ($\mu g\ g^{-1}$) in household dust from different countries.

PAHs	This Study								[91]	[85]	[87]		[88]		[92]			
	Kitchens		Living Rooms		Bedrooms		Rug Bedroom	Rug Living Room	House Vacuum Samples Brisbane	Indoor Dust from Distinct Buildings Nepal	Settled Dust from Homes in Ottawa		Dust Samples from Private Domiciles and Public Buildings, China		Saudi Household Floor Dust		Kuwaiti Household Floor Dust	
	Mean Min-Max	Median	Mean Min-Max	Median	Mean Min-Max	Median	Mean	Mean	Mean	Median	Mean	Median	Mean	Median	Mean	Median	Mean	Median
ACY	1.75 <0.001–7.89	0.432	3.22 <0.001–11.8	1.15	4.70 0.125–13.0	2.04	0.988	<0.001	6.92	0.031	0.039	0.005	0.173	0.055	0.160	0.080	0.090	0.060
ACE	0.344 <0.001–1.50	<0.001	0.74 <0.001–2.81	0.215	0.246 <0.001–1.45	0.063	0.067	0.136	202	0.018			0.093	0.052	0.110	0.105	0.065	0.060
FLU	0.073 <0.001–0.349	<0.001	0.036 <0.001–0.287	<0.001	0.017 <0.001–0.081	<0.001	0.036	0.012	12.6	0.048	0.170	0.093	0.464	0.200	0.290	0.170	0.165	0.080
PHE	0.213 <0.001–0.876	<0.001	0.260 <0.001–0.987	0.154	0.306 <0.001–1.16	0.177	0.169	0.275	8.77	0.173	2.78	1.48	4.19	1.67	0.425	0.160	0.510	0.410
ANT	1.56 <0.001–10.0	<0.001	1.07 <0.001–8.57	<0.001	< 0.001 <0.001–0.003	<0.001	0.002	1.52	3.61	0.062	0.485	0.196	0.418	0.196	0.105	0.050	0.032	0.025
FLUA	1.56 <0.001–8.48	0.277	1.12 <0.001–4.58	<0.001	1.60 <0.001–11.4	0.045	<0.001	0.197	53.6	0.128			4.26	1.77	0.425	0.175	0.220	0.140
PYR	4.46 <0.001–19.3	2.17	3.67 <0.001–11.2	1.51	4.65 <0.001–27.0	0.604	0.003	1.06	77.3	0.107			2.89	1.18	0.385	0.160	0.035	0.017
CHR	0.921 <0.001–3.65	0.404	0.695 <0.001–2.36	0.506	0.610 <0.001–3.41	0.23	0.120	0.053	5.42	0.157	3.29	1.19	1.79	0.717	0.095	0.085	0.055	0.040
PER	4.29 <0.001–22.4	<0.001	3.73 <0.001–16.8	1.70	3.26 0.001–14.3	1.25	0.301	0.602										
CAR	1.08 <0.001–7.89	<0.001	1.83 <0.001–7.76	0.440	1.64 <0.001–11.6	<0.001	0.001	0.141										
TER	<0.001	<0.001	0.135 <0.001–0.765	<0.001	0.021 <0.001–0.106	<0.001	0.028	<0.001										
RET	3.15 <0.001–15.4	0.737	5.02 <0.001–16.2	2.20	5.28 <0.001–21.4	0.966	<0.001	2.09										
BaA	1.55 <0.001–3.60	1.78	0.516 <0.001–1.67	0.136	0.333 <0.001–2.38	<0.001	0.040	1.40	6.39	0.078	2.38	0.696	2.37	0.569	0.090	0.080	0.090	0.055

Table 3. *Cont.*

PAHs	This Study								[91]	[85]	[87]		[88]		[92]			
	Kitchens		Living Rooms		Bedrooms		Rug Bedroom	Rug Living Room	House Vacuum Samples Brisbane	Indoor Dust from Distinct Buildings Nepal	Settled Dust from Homes in Ottawa		Dust Samples from Private Domiciles and Public Buildings, China		Saudi Household Floor Dust		Kuwaiti Household Floor Dust	
	Mean Min-Max	Median	Mean Min-Max	Median	Mean Min-Max	Median	Mean	Mean	Mean	Median	Mean	Median	Mean	Median	Mean	Median	Mean	Median
BbF	1.91 <0.001–6.68	1.63	2.10 <0.001–5.28	1.31	1.50 <0.001–6.21	0.124	0.304	1.07	11.3	0.018	4.87	1.66	6.23	1.52	0.700	0.575	0.245	0.210
BkF	1.10 <0.001–5.56	<0.001	1.02 <0.001–3.51	< 0.001	0.960 <0.001–4.84	<0.001	<0.001	0.710	3.48	0.075	1.60	0.582	0.763	0.248	0.110	0.105	0.070	0.040
BeP	0.206 <0.001–1.44	<0.001	0.844 <0.001–4.50	<0.001	<0.001	<0.001	<0.001	0.746										
BaP	1.53 <0.001–4.91	1.63	1.23 <0.001–3.27	0.564	1.11 <0.001–3.43	0.219	0.061	1.43	2.97	0.094	2.91	0.803	1.71	0.310	0.550	0.290	0.140	0.080
IcdP	3.48 0.170–12.9	1.75	3.11 0.979–5.13	2.85	2.76 <0.001–8.29	2.32	2.61	0.565	3.34	0.115	3.07	0.911	1.97	0.356	0.015	<0.001	0.055	<0.001
DahA	0.289 0.290–1.22	<0.001	0.790 <0.001–4.31	0.314	0.395 <0.001–1.52	0.155	0.162	0.071	20.1	0.046	0549	0.185	0.510	0.090	0.016	<0.001	0.015	<0.001
BghiP	2.43 <0.001–13.2	0.922	1.10 <0.001–4.40	0.275	1.46 <0.001–6.40	<0.001	<0.001	0.263	6.08	0.068	2.79	0.793	2.14	0.398	0.105	0.075	0.070	<0.001

ACY, acenaphthylene; ACE, acenaphthene; FLU, fluorene; PHE, phenanthrene; ANT, anthracene; FLUA, fluoranthene; PYR, pyrene; CHR, chrysene; PER, perylene; CAR, carbazole; TER, p-terphenyl; RET, retene; BaA, benzo[a]anthracene; BbF, benzo[b]fluoranthene; BkF, benzo[k]fluoranthene; BeP, benzo[e]pyrene; BaP, benzo[a]pyrene; IcdP, indeno[1,2,3-cd]pyrene; DahA, dibenzo[a,h]anthracene; and BghiP, benzo[ghi]perylene.

3.3. Human Exposure to Plasticizers and PAHs in Resuspended Dust

Ingestion is the main pathway for intake of plasticizers from dust (Table 4). Regardless of the route, household residents are exposed to higher intakes in the bedrooms, whereas the lowest doses are experienced in the kitchens. The daily intakes for children and adults (13 to 29 and 1.4 to 3.1 μg kg^{-1} day^{-1}, respectively) via dust ingestion were three to four orders of magnitude higher than those via inhalation (0.357 to 0.802 and 0.129 to 0.289 ng kg^{-1} day^{-1}, respectively) and dermal contact (14.9 to 33.0 and 2.05 to 4.62 ng kg^{-1} day^{-1}, respectively). Children are at higher risk of exposure to plasticizers than adults. The total DI_{ing}, DI_{inh}, and DI_{der} for children were about 9.3, 2.8, and 7.1 times higher than those estimated for adults, respectively. DBP, DNOP, and DEHP contributed the most to daily intakes. The dust ingestion intakes for these compounds were lower than the U.S. EPA maximum acceptable oral doses of 0.1, 0.01, and 0.02 mg kg^{-1} day^{-1}, respectively. However, the daily intakes of plasticizers through dust ingestion, inhalation, and dermal contact in this study are higher than those estimated for indoor dust from houses of several Chinese regions [39], and childcare facilities, salons, and homes across the USA [29]. Albar et al. [37] assessed human exposure to phthalates via dust ingestion for the worst-case scenario (with 95th percentile levels) for Saudi and Kuwaiti toddlers and adults. The exposure to DEHP, which is cardiotoxic and endocrine disruptor, for Saudi toddlers was estimated to be 37,630 ng kg^{-1} day^{-1}, while for Kuwaiti toddlers it was 6722 ng kg^{-1} day^{-1}. Similarly, exposure estimates to other PAEs, such as DBP, DIBP, and DNOP, was also higher for Saudi toddlers. In the case of Saudi and Kuwaiti adults, dust ingestion intakes for DEHP were estimated at 1613 and 288 ng kg^{-1} day^{-1}, respectively.

Table 4. Mean daily intakes (ng kg^{-1} day^{-1}) of plasticizers via dust ingestion, inhalation, and dermal contact for children and adults.

Room	DMP	DEP	DBP	BBP	DEHA	DEHP	DNOP	DINP	DIDP	Σ
				DI_{ing}–CHILDREN						
Kitchens	1.78	934	3695	41.4	112	1782	6095	67.1	43.4	12,772
Bedrooms	2.17	296	15,472	68.4	237	5898	6372	131	232	28,707
Living rooms	6.12	914	7765	73.0	125	8890	4938	185	437	23,333
				DI_{inh}–CHILDREN						
Kitchens	0.000	0.0261	0.103	0.003	0.001	0.050	0.170	0.002	0.001	0.357
Bedrooms	0.000	0.0083	0.432	0.007	0.002	0.165	0.178	0.004	0.006	0.802
Living rooms	0.000	0.0255	0.217	0.003	0.002	0.248	0.138	0.005	0.012	0.652
				D_{Ider}–CHILDREN						
Kitchens	0.002	1.07	4.25	0.129	0.048	2.05	7.01	0.077	0.050	14.7
Bedrooms	0.003	0.340	17.8	0.272	0.079	6.78	7.33	0.150	0.267	33.0
Living rooms	0.007	1.05	8.93	0.144	0.084	10.2	5.68	0.213	0.502	26.8
				DI_{ing}–ADULTS						
Kitchens	0.190	100	396	12.0	4.44	191	653	7.19	4.65	1368
Bedrooms	0.232	31.7	1658	25.4	7.33	632	683	14.0	24.9	3076
Living rooms	0.655	97.9	832	13.4	7.82	952	529	19.9	46.8	2500
				DI_{inh}–ADULTS						
Kitchens	0.000	0.009	0.037	0.001	0.000	0.018	0.062	0.001	0.000	0.129
Bedrooms	0.000	0.003	0.156	0.002	0.001	0.059	0.064	0.001	0.002	0.289
Living rooms	0.000	0.009	0.078	0.001	0.001	0.090	0.050	0.002	0.004	0.235
				DI_{der}–ADULTS						
Kitchens	0.000	0.150	0.595	0.018	0.0067	0.287	0.981	0.011	0.007	2.05
Bedrooms	0.000	0.048	2.49	0.038	0.0110	0.949	1.03	0.021	0.037	4.62
Living rooms	0.001	0.147	1.25	0.020	0.0117	1.43	0.794	0.030	0.070	3.75

Reference doses for oral exposure, R_fD (mg kg^{-1} day^{-1}) recommended by the United States Environmental Protection Agency: DEP, 0.8; DBP, 0.1; BBP, 0.2; DEHA, 0.6; DEHP, 0.02; and DNOP, 0.01. Reference doses for other compounds or other exposure pathways are not available.

BaP, DahA, and IcdP were the compounds that most contributed to the carcinogenic potency, accounting for 45% to 57%, 11% to 29% and 11% to 13% of the total BaP_{TEQ}, respectively. As shown in Table 5, the total cancer risk could be attributed almost entirely to ingestion and did not vary much with the microenvironment. Therefore, inhalation of resuspended particles through the mouth and nose or via dermal contact was almost negligible as compared with the ingestion route. Under most regulatory programs, an ILCR between 10^{-6} and 10^{-4} indicates potential risk, an ILCR of 10^{-6} or less is considered insignificant, and an ILCR $\geq 10^{-4}$ is taken as high risk [93,94]. In the present study, the total risk of adult and children exposure to PAHs in dust via the three pathways ranged from 7.2×10^{-6} to 1.4×10^{-5}. This means that the resuspendable thoracic fraction of household dust can contribute to an estimated excess of 7.2 to 14 per million people new cancer cases. One cancer case per million people is usually used as a baseline level of acceptable risk.

Table 5. Incremental lifetime cancer risk from human exposure to PAHs in PM_{10} resuspended from household dust via ingestion, inhalation, and dermal absorption.

Cancer risk		Kitchens	Bedrooms	Living Rooms
$ILCR_{ing}$	Children	9.2×10^{-6}	7.2×10^{-6}	9.4×10^{-6}
	Adults	1.4×10^{-5}	1.1×10^{-5}	1.4×10^{-5}
$ILCR_{inh}$	Children	1.4×10^{-10}	1.1×10^{-10}	1.4×10^{-10}
	Adults	6.9×10^{-10}	5.4×10^{-10}	9.0×10^{-10}
$ILCR_{der}$	Children	3.6×10^{-8}	2.8×10^{-8}	3.7×10^{-8}
	Adults	7.1×10^{-8}	5.6×10^{-8}	7.3×10^{-8}
Total ILCR	Children	9.3×10^{-6}	7.2×10^{-6}	9.4×10^{-6}
	Adults	1.4×10^{-5}	1.1×10^{-5}	1.4×10^{-5}

To assess the potential health risk of foliar dust in Nanjing, China, Zha et al. [56] analyzed the contents of 16 priority PAHs. Total concentrations in dust ranged from 1.77 to 19.02 $\mu g\ g^{-1}$, with an average value of 6.98 $\mu g\ g^{-1}$. The cancer risk levels via dermal contact and ingestion varied from 10^{-8} to 10^{-6} in all the dust samples, while the mean cancer risk via inhalation was 10^{-10} to 10^{-12}, about 10^4 to 10^7 times lower than through ingestion and dermal contact.

ILCR values due to human exposure to PAHs in indoor dust in city, town, village, and orefield of Guizhou province, China, were 6.14×10^{-6}, 5.00×10^{-6}, 3.08×10^{-6}, and 6.02×10^{-6} for children and 5.92×10^{-6}, 4.83×10^{-6}, 2.97×10^{-6}, and 5.81×10^{-6} for adults, respectively [16]. As noted in our study, inhalation of resuspended particles through mouth and nose was very small as compared with the ingestion pathway. Maertens et al. [87] estimated higher excess lifetime cancer risks, ranging from 1×10^{-6} and 1×10^{-4}, associated with nondietary ingestion of PAHs in settled dust from homes in Ottawa, Canada, during preschool years. However, the researchers observed that the level of risk varies substantially according to the ingestion rates adopted to perform the estimates.

Since ingestion was found to be the dominant exposure pathway, the risk associated with seven non-carcinogenic PAHs (ACY, ACE, FLU, PHE, ANT, FLUA, and PYR) in household PM_{10} dust was estimated through the hazard quotient (HQ). Under most programs, if the HQ value is greater than one, the exposed population is likely to experience considerable non-carcinogenic effects. The highest daily intakes were obtained for PYR (24 to 31 ng kg^{-1} day^{-1} for children and 2.6 to 3.3 ng kg^{-1} day^{-1} for adults) and ACY (12 to 31 ng kg^{-1} day^{-1} for children and 1.2 to 3.3 ng kg^{-1} day^{-1} for adults). These values are higher than those documented for dust samples collected from three different microenvironments (cars, air conditioner filters, and household floor dust) of Jeddah, Saudi Arabia, and Kuwait [92]. Nevertheless, the HQ values of our study were less than one, indicating that there was no considerable non-carcinogenic risk arising from ingestion of PAHs in resuspended PM_{10}. Although the risk remains very low for both age groups, higher HQ values for children (1.51×10^{-3} to 1.89×10^{-3}) than those obtained for adults (1.32×10^{-4} to 2.03×10^{-4}), demonstrate greater susceptibility of the younger ones. It should be noted that HQ values are underestimated as naphthalene was not included in the quantifications. Despite its volatility, naphthalene is normally quite abundant in the particulate phase.

4. Conclusions

This preliminary study provides a first insight on the occurrence of plasticizers and PAHs in PM_{10} from resuspended dust samples in Spanish households and adds to the growing evidence that non-dietary exposure contributes to the total body burden. Considering that people spend most of their time indoors, exposure to these pollutants could lead to an increased human health risk. Although no appreciable differences between plasticizer and PAH levels in resuspended dust from the various residential microenvironments were observed, it was concluded that exposure through the ingestion route poses much higher risks as compared to inhalation and dermal contact. This is of particular concern for infants due to their higher dust intake via frequent hand-to-mouth activities. Because of the small number of samples analyzed in this study, it should be noted that exposure estimates are only an indication of the likely range for children and adults within the studied population. More assessments with wider spatial and temporal coverage are needed to better understand the dynamics and possible effects of these pollutants in different indoor microenvironments. As observed in other works, this study on exposure to organic pollutants suggests that measurements only in food and outdoor environments can substantially underestimate exposures to chemicals. Comparisons of the results of this study with those reported in the literature revealed huge amplitudes in the numbers and great difficulty in generalizing conclusions, mainly due to the heterogeneity of applied methodologies. Traditional methodologies have been compartmentalized (settled versus suspended dust) and as a result may have described the environment incompletely. Thus, the scientific community and international agencies should discuss and establish standardized protocols.

Supplementary Materials: The following are available online at http://www.mdpi.com/2073-4433/10/12/785/s1, Analytical technique for quantification of the carbonaceous content of PM_{10}, Table S1: Standards from Sigma-Aldrich (and product references) used in chromatographic analysis, Table S2: Dust loading obtained in the various houses, in both sampling periods, mass percentage of carbonaceous material (TC = OC + EC) and mass fractions of plasticizers and polycyclic aromatic hydrocarbons in PM_{10}.

Author Contributions: Conceptualization, C.A., E.D.V., and F.A.; sampling, E.D.V., A.C. (A. Calvo), C.d.B.-A., F.O., A.C. (A. Castro), and R.F.; carbon analyzes, T.N.; filter solvent extractions, E.D.V.; GC-MS analyzes, A.V.; writing of original draft, C.A.; review of original draft, C.A., in collaboration with all co-authors.

Funding: The sampling and analytical work was mainly supported by the project POCI-01-0145-FEDER-029574 (SOPRO- Chemical and toxicological SOurce PROfiling of particulate matter in urban air), funded by FEDER, through COMPETE2020 - Programa Operacional Competitividade e Internacionalização (POCI), and by national funds (OE), through FCT/MCTES. The authors would like to express their gratitude to the Portuguese Foundation of Science and Technology (FCT) and to the POHP/FSE funding program for the fellowship SFRH/BD/117993/2016. Ana Vicente was supported by national funds (OE), through FCT, I.P., in the scope of the framework contract foreseen in the numbers 4, 5, and 6 of article 23, of the Decree-Law 57/2016, of August 29, changed by Law 57/2017, of July 19. Some support was received from CESAM (UID/AMB/50017), which was funded by FCT/MEC through national funds, and co-funded by FEDER, within the PT2020 Partnership Agreement and Compete 2020.

Acknowledgments: We thank the homeowners who allowed sampling in various rooms.

Conflicts of Interest: The authors declare no conflict of interest.

References

1. Schweizer, C.; Edwards, R.D.; Bayer-Oglesby, L.; Gauderman, W.J.; Ilacqua, V.; Juhani Jantunen, M.; Lai, H.K.; Nieuwenhuijsen, M.; Künzli, N. Indoor time–microenvironment–activity patterns in seven regions of Europe. *J. Expo. Sci. Environ. Epidemiol.* **2007**, *17*, 170–181. [CrossRef] [PubMed]
2. Qian, J.; Peccia, J.; Ferro, A.R. Walking-induced particle resuspension in indoor environments. *Atmos. Environ.* **2014**, *89*, 464–481. [CrossRef]
3. Tian, Y.; Sul, K.; Qian, J.; Mondal, S.; Ferro, A.R. A comparative study of walking-induced dust resuspension using a consistent test mechanism. *Indoor Air* **2014**, *24*, 592–603. [CrossRef] [PubMed]

4. Bo, M.; Salizzoni, P.; Clerico, M.; Buccolieri, R. Assessment of indoor-outdoor particulate matter air pollution: A review. *Atmosphere* **2017**, *8*, 136. [CrossRef]
5. Naspinski, C.; Lingenfelter, R.; He, L.Y.; Cizmas, L.; Naufal, Z.; Islamzadeh, A.; Li, Z.; Li, Z.; Donnelly, K.C.; McDonald, T. A comparison of concentrations of polycyclic aromatic compounds detected in dust samples from various regions of the world. *Environ. Int.* **2008**, *34*, 988–993. [CrossRef]
6. Lioy, P.J.; Freeman, N.C.G.; Millette, J.R. Review Dust: A Metric for Use in Residential and Building Exposure Assessment and. *Environ. Health Perspect.* **2002**, *110*, 969–983. [CrossRef]
7. Maertens, R.M.; Bailey, J.; White, P.A. The mutagenic hazards of settled house dust: A review. *Mutat. Res.* **2004**, *567*, 401–425. [CrossRef]
8. Liu, H.Y.; Dunea, D.; Iordache, S.; Pohoata, A. A review of airborne particulate matter effects on young children's respiratory symptoms and diseases. *Atmosphere* **2018**, *9*, 150. [CrossRef]
9. Cipriani, F.; Calamelli, E.; Ricci, G. Allergen avoidance in allergic asthma. *Front. Pediatr.* **2017**, *5*, 103. [CrossRef]
10. Ahluwalia, S.K.; Matsui, E.C. The indoor environment and its effects on childhood asthma. *Curr. Opin. Allergy Clin. Immunol.* **2011**, *11*, 137–143. [CrossRef]
11. McCormack, M.C.; Breysse, P.N.; Matsui, E.C.; Hansel, N.N.; Peng, R.D.; Curtin-Brosnan, J.; Williams, D.L.; Wills-Karp, M.; Diette, G.B.; Center for Childhood Asthma in the Urban Environment. Indoor particulate matter increases asthma morbidity in children with non-atopic and atopic asthma. *Ann. Allergy. Asthma Immunol.* **2011**, *106*, 308–315. [CrossRef] [PubMed]
12. Mitro, S.D.; Dodson, R.E.; Singla, V.; Adamkiewicz, G.; Elmi, A.F.; Tilly, M.K.; Zota, A.R. Consumer Product Chemicals in Indoor Dust: A Quantitative Meta-analysis of U.S. Studies. *Environ. Sci. Technol.* **2016**, *50*, 10661–10672. [CrossRef] [PubMed]
13. Roberts, J.W.; Wallace, L.A.; Camann, D.E.; Dickey, P.; Gilbert, S.G.; Lewis, R.G.; Takaro, T.K. Monitoring and Reducing Exposure of Infants to Pollutants in House Dust. In *Reviews of Environmental Contamination and Toxicology*; Springer: Berlin, Germany, 2009; Volume 201, pp. 1–39.
14. Langer, S.; Weschler, C.J.; Fischer, A.; Bekö, G.; Toftum, J.; Clausen, G. Phthalate and PAH concentrations in dust collected from Danish homes and daycare centers. *Atmos. Environ.* **2010**, *44*, 2294–2301. [CrossRef]
15. Dodson, R.E.; Camann, D.E.; Morello-Frosch, R.; Brody, J.G.; Rudel, R.A. Semivolatile organic compounds in homes: strategies for efficient and systematic exposure measurement based on empirical and theoretical factors. *Environ. Sci. Technol.* **2015**, *49*, 113–122. [CrossRef] [PubMed]
16. Yang, Q.; Chen, H.; Li, B. Polycyclic aromatic hydrocarbons (PAHs) in indoor dusts of Guizhou, southwest of China: Status, sources and potential human health risk. *PLoS ONE* **2015**, *10*, e0118141. [CrossRef]
17. Mercier, F.; Glorennec, P.; Thomas, O.; Bot, B. Le Organic contamination of settled house dust, a review for exposure assessment purposes. *Environ. Sci. Technol.* **2011**, *45*, 6716–6727. [CrossRef]
18. DellaValle, C.T.; Deziel, N.C.; Jones, R.R.; Colt, J.S.; De Roos, A.J.; Cerhan, J.R.; Cozen, W.; Severson, R.K.; Flory, A.R.; Morton, L.M.; et al. Polycyclic aromatic hydrocarbons: Determinants of residential carpet dust levels and risk of non-Hodgkin lymphoma. *Cancer Causes Control* **2016**, *27*, 1–13. [CrossRef]
19. Boström, C.E.; Gerde, P.; Hanberg, A.; Jernström, B.; Johansson, C.; Kyrklund, T.; Rannug, A.; Törnqvist, M.; Victorin, K.; Westerholm, R. Cancer risk assessment, indicators, and guidelines for polycyclic aromatic hydrocarbons in the ambient air. *Environ. Health Perspect.* **2002**, *110*, 451–488.
20. Kim, K.H.; Jahan, S.A.; Kabir, E.; Brown, R.J.C. A review of airborne polycyclic aromatic hydrocarbons (PAHs) and their human health effects. *Environ. Int.* **2013**, *60*, 71–80. [CrossRef]
21. Deziel, N.C.; Rull, R.P.; Colt, J.S.; Reynolds, P.; Whitehead, T.P.; Gunier, R.B.; Month, S.R.; Taggart, D.R.; Buffler, P.; Ward, M.H.; et al. Polycyclic aromatic hydrocarbons in residential dust and risk of childhood acute lymphoblastic leukemia. *Environ. Res.* **2014**, *133*, 388–395. [CrossRef]
22. Wang, B.L.; Pang, S.T.; Zhang, X.L.; Li, X.L.; Sun, Y.G.; Lu, X.M.; Zhang, Q.; Zhang, Z.D. Levels and neurodevelopmental effects of polycyclic aromatic hydrocarbons in settled house dust of urban dwellings on preschool-aged children in Nanjing, China. *Atmos. Pollut. Res.* **2014**, *5*, 292–302. [CrossRef]
23. Kang, Y.; Cheung, K.C.; Wong, M.H. Polycyclic aromatic hydrocarbons (PAHs) in different indoor dusts and their potential cytotoxicity based on two human cell lines. *Environ. Int.* **2010**, *36*, 542–547. [CrossRef] [PubMed]
24. Liu, R.; He, R.; Cui, X.; Ma, L.Q. Impact of particle size on distribution, bioaccessibility, and cytotoxicity of polycyclic aromatic hydrocarbons in indoor dust. *J. Hazard. Mater.* **2018**, *357*, 341–347. [CrossRef] [PubMed]

25. World Health Organization International Agency for Research on Cancer. *IARC Monographs on the Evaluation of Carcinogenic Risks to Humans. Some Non-Heterocyclic Polycyclic Aromatic Hydrocarbons and Some Related Exposures*; WHO: Lyon, France, 2010.
26. ATSDR. Priority List of Hazardous Substances. Available online: http://www.atsdr.cdc.gov/SPL/resources/ (accessed on 5 December 2019).
27. Ma, W.-L.; Subedi, B.; Kannan, K. The Occurrence of Bisphenol A, Phthalates, Parabens and Other Environmental Phenolic Compounds in House Dust: A Review. *Curr. Org. Chem.* **2014**, *18*, 2182–2199. [CrossRef]
28. Larsson, K.; Lindh, C.H.; Jönsson, B.A.; Giovanoulis, G.; Bibi, M.; Bottai, M.; Bergström, A.; Berglund, M. Phthalates, non-phthalate plasticizers and bisphenols in Swedish preschool dust in relation to children's exposure. *Environ. Int.* **2017**, *102*, 114–124. [CrossRef]
29. Subedi, B.; Sullivan, K.D.; Dhungana, B. Phthalate and non-phthalate plasticizers in indoor dust from childcare facilities, salons, and homes across the USA. *Environ. Pollut.* **2017**, *230*, 701–708. [CrossRef]
30. Kang, Y.; Man, Y.B.; Cheung, K.C.; Wong, M.H. Risk assessment of human exposure to bioaccessible phthalate esters via indoor dust around the Pearl River Delta. *Environ. Sci. Technol.* **2012**, *46*, 8422–8430. [CrossRef]
31. Kweon, D.J.; Kim, M.K.; Zoh, K.D. Distribution of brominated flame retardants and phthalate esters in house dust in Korea. *Environ. Eng. Res.* **2018**, *23*, 354–363. [CrossRef]
32. Abb, M.; Heinrich, T.; Sorkau, E.; Lorenz, W. Phthalates in house dust. *Environ. Int.* **2009**, *35*, 965–970. [CrossRef]
33. Kubwabo, C.; Rasmussen, P.E.; Fan, X.; Kosarac, I.; Wu, F.; Zidek, A.; Kuchta, S.L. Analysis of selected phthalates in Canadian indoor dust collected using household vacuum and standardized sampling techniques. *Indoor Air* **2013**, *23*, 506–514. [CrossRef]
34. Bornehag, C.G.; Lundgren, B.; Weschler, C.J.; Sigsgaard, T.; Hagerhed-Engman, L.; Sundell, J. Phthalates in indoor dust and their association with building characteristics. *Environ. Health Perspect.* **2005**, *113*, 1399–1404. [CrossRef] [PubMed]
35. Al Qasmi, N.N.; Al-Thaiban, H.; Helaleh, M.I.H. Indoor phthalates from household dust in Qatar: Implications for non-dietary human exposure. *Environ. Sci. Pollut. Res. Int.* **2019**, *26*, 421–430. [CrossRef] [PubMed]
36. Zhang, Q.; Lu, X.M.; Zhang, X.L.; Sun, Y.G.; Zhu, D.M.; Wang, B.L.; Zhao, R.Z.; Zhang, Z.D. Levels of phthalate esters in settled house dust from urban dwellings with young children in Nanjing, China. *Atmos. Environ.* **2013**, *69*, 258–264. [CrossRef]
37. Albar, H.M.S.A.; Ali, N.; Shahzad, K.; Ismail, I.M.I.; Rashid, M.I.; Wang, W.; Ali, L.N.; Eqani, S.A.M.A.S. Phthalate esters in settled dust of different indoor microenvironments; source of non-dietary human exposure. *Microchem. J.* **2017**, *132*, 227–232.
38. Kashyap, D.; Agarwal, T. Concentration and factors affecting the distribution of phthalates in the air and dust: A global scenario. *Sci. Total Environ.* **2018**, *635*, 817–827. [CrossRef]
39. Zhu, Q.; Jia, J.; Zhang, K.; Zhang, H.; Liao, C.; Jiang, G. Phthalate esters in indoor dust from several regions, China and their implications for human exposure. *Sci. Total Environ.* **2019**, *652*, 1187–1194. [CrossRef]
40. Guo, Y.; Kannan, K. Comparative assessment of human exposure to phthalate esters from house dust in China and the United States. *Environ. Sci. Technol.* **2011**, *45*, 3788–3794. [CrossRef]
41. Orecchio, S.; Indelicato, R.; Barreca, S. The distribution of phthalate esters in indoor dust of Palermo (Italy). *Environ. Geochem. Health* **2013**, *35*, 613–624. [CrossRef]
42. Kay, V.R.; Bloom, M.S.; Foster, W.G. Reproductive and developmental effects of phthalate diesters in males. *Crit. Rev. Toxicol.* **2014**, *44*, 467–498. [CrossRef]
43. Kay, V.R.; Chambers, C.; Foster, W.G. Reproductive and developmental effects of phthalate diesters in females. *Crit. Rev. Toxicol.* **2013**, *43*, 200–219. [CrossRef]
44. Wormuth, M.; Scheringer, M.; Vollenweider, M.; Hungerbühler, K. What are the sources of exposure to eight frequently used phthalic acid esters in Europeans? *Risk Anal.* **2006**, *26*, 803–824. [CrossRef] [PubMed]
45. Directive 2005/84/EC of the European Parliament and of the Council of 14 December 2005 amending for the 22nd time Council Directive 76/769/EEC on the approximation of the laws, regulations and administrative provisions of the Member States relating to restrictions on the marketing and use of certain dangerous substances and preparations (phthalates in toys and childcare articles). Available online: https://eur-lex.europa.eu/LexUriServ/LexUriServ.do?uri=OJ:L:2005:344:0040:0043:EN:PDF (accessed on 5 December 2019).

46. Manigrasso, M.; Guerriero, E.; Avino, P. Ultrafine particles in residential indoors and doses deposited in the human respiratory system. *Atmosphere* **2015**, *6*, 1444–1461. [CrossRef]
47. Morawska, L.; Salthammer, T. (Eds.) *Indoor Environment, Airborne Particles and Settled Dust*; Wiley-VCH GmbH & Co. KGaA: Weinheim, Germany, 2003; ISBN 978-352-7305254.
48. Ma, Y.; Harrad, S. Spatiotemporal analysis and human exposure assessment on polycyclic aromatic hydrocarbons in indoor air, settled house dust, and diet: A review. *Environ. Int.* **2015**, *84*, 7–16. [CrossRef] [PubMed]
49. Driver, J.H.; Konz, J.J.; Whitmyre, G.K. Soil adherence to human skin. *Bull. Environ. Contam. Toxicol.* **1989**, *43*, 814–820. [CrossRef] [PubMed]
50. Amato, F.; Pandolfi, M.; Moreno, T.; Furger, M.; Pey, J.; Alastuey, A.; Bukowiecki, N.; Prevot, A.S.H.; Baltensperger, U.; Querol, X. Sources and variability of inhalable road dust particles in three European cities. *Atmos. Environ.* **2011**, *45*, 6777–6787. [CrossRef]
51. Amato, F.; Pandolfi, M.; Viana, M.; Querol, X.; Alastuey, A.; Moreno, T. Spatial and chemical patterns of PM10 in road dust deposited in urban environment. *Atmos. Environ.* **2009**, *43*, 1650–1659. [CrossRef]
52. Alves, C.A.; Evtyugina, M.; Vicente, A.M.P.; Vicente, E.D.; Nunes, T.V.; Silva, P.M.A.; Duarte, M.A.C.; Pio, C.A.; Amato, F.; Querol, X. Chemical profiling of PM_{10} from urban road dust. *Sci. Total Environ.* **2018**, *634*, 41–51. [CrossRef]
53. Anderson, J.O.; Thundiyil, J.G.; Stolbach, A. Clearing the Air: A Review of the Effects of Particulate Matter Air Pollution on Human Health. *J. Med. Toxicol.* **2012**, *8*, 166–175. [CrossRef]
54. Alves, C.A.; Vicente, A.; Monteiro, C.; Gonçalves, C.; Evtyugina, M.; Pio, C. Emission of trace gases and organic components in smoke particles from a wildfire in a mixed-evergreen forest in Portugal. *Sci. Total Environ.* **2011**, *409*, 1466–1475. [CrossRef]
55. Fortune, A.; Gendron, L.; Tuday, M. Comparison of naphthalene ambient air sampling & analysis methods at former manufactured gas plant (MGP) remediation sites. *Int. J. Soil Sediment Water* **2010**, *3*, 1–15.
56. Zha, Y.; Zhang, Y.L.; Tang, J.; Sun, K. Status, sources, and human health risk assessment of PAHs via foliar dust from different functional areas in Nanjing, China. *J. Environ. Sci. Health Part A* **2018**, *53*, 571–582. [CrossRef] [PubMed]
57. Iwegbue, C.M.A.; Obi, G.; Uzoekwe, S.A.; Egobueze, F.E.; Odali, E.W.; Tesi, G.O.; Nwajei, G.E.; Martincigh, B.S. Distribution, sources and risk of exposure to polycyclic aromatic hydrocarbons in indoor dusts from electronic repair workshops in southern Nigeria. *Emerg. Contam.* **2019**, *5*, 23–30. [CrossRef]
58. Abdullahi, K.L.; Delgado-Saborit, J.M.; Harrison, R.M. Emissions and indoor concentrations of particulate matter and its specific chemical components from cooking: A review. *Atmos. Environ.* **2013**, *71*, 260–294. [CrossRef]
59. Licina, D.; Tian, Y.; Nazaroff, W.W. Emission rates and the personal cloud effect associated with particle release from the perihuman environment. *Indoor Air* **2017**, *27*, 791–802. [CrossRef] [PubMed]
60. Lewis, R.D.; Ong, K.H.; Emo, B.; Kennedy, J.; Kesavan, J.; Elliot, M. Resuspension of house dust and allergens during walking and vacuum cleaning. *J. Occup. Environ. Hyg.* **2018**, *15*, 235–245. [CrossRef]
61. Becher, R.; Øvrevik, J.; Schwarze, P.E.; Nilsen, S.; Hongslo, J.K.; Bakke, J.V. Do carpets impair indoor air quality and cause adverse health outcomes: A review. *Int. J. Environ. Res. Public Health* **2018**, *15*, 184. [CrossRef]
62. Foarde, K.; Berry, M. Comparison of biocontaminant levels associated with hard vs. carpet floors in nonproblem schools: Results of a year long study. *J. Expo. Anal. Environ. Epidemiol.* **2004**, *14*, S41–S48. [CrossRef]
63. Tranter, D.C. Indoor allergens in settled school dust: A review of findings and significant factors. *Clin. Exp. Allergy* **2005**, *35*, 126–136. [CrossRef]
64. Causer, S.; Shorter, C.; Sercombe, J. Effect of floorcovering construction on content and vertical distribution of house dust mite allergen, Der p I. *J. Occup. Environ. Hyg.* **2006**, *3*, 161–168. [CrossRef]
65. Roberts, J.W.; Clifford, W.S.; Glass, G.; Hummer, P.G. Reducing dust, lead, dust mites, bacteria, and fungi in carpets by vacuuming. *Arch. Environ. Contam. Toxicol.* **1999**, *36*, 477–484.
66. Adgate, J.L.; Weisel, C.; Wang, Y.; Rhoads, G.G.; Lioy, P.J. Lead in house dust: Relationships between exposure metrics. *Environ. Res.* **1995**, *70*, 134–147. [CrossRef] [PubMed]

67. Lioy, P.J.; Freeman, N.C.; Wainman, T.; Stern, A.H.; Boesch, R.; Howell, T.; Shupack, S.I. Microenvironmental analysis of residential exposure to chromium-laden wastes in and around New Jersey homes. *Risk Anal.* **1992**, *12*, 287–299. [CrossRef] [PubMed]
68. Bai, Z.; Yiin, L.M.; Rich, D.Q.; Adgate, J.L.; Ashley, P.J.; Lioy, P.J.; Rhoads, G.G.; Zhang, J. Field evaluation and comparison of five methods of sampling lead dust on carpets. *Am. Ind. Hyg. Assoc. J.* **2003**, *64*, 528–532. [CrossRef]
69. Polidori, A.; Turpin, B.; Meng, Q.Y.; Lee, J.H.; Weisel, C.; Morandi, M.; Colome, S.; Stock, T.; Winer, A.; Zhang, J.; et al. Fine organic particulate matter dominates indoor-generated PM2.5 in RIOPA homes. *J. Expo. Sci. Environ. Epidemiol.* **2006**, *16*, 321–331. [CrossRef] [PubMed]
70. Bamai, Y.A.; Araki, A.; Kawai, T.; Tsuboi, T.; Saito, I.; Yoshioka, E.; Kanazawa, A.; Tajima, S.; Shi, C.; Tamakoshi, A.; et al. Associations of phthalate concentrations in floor dust and multi-surface dust with the interior materials in Japanese dwellings. *Sci. Total Environ.* **2014**, *468*, 147–157. [CrossRef]
71. Afshari, A.; Gunnarsen, L.; Clausen, P.A.; Hansen, V. Emission of phthalates from PVC and other materials. *Indoor Air* **2004**, *14*, 120–128. [CrossRef]
72. Koniecki, D.; Wang, R.; Moody, R.P.; Zhu, J. Phthalates in cosmetic and personal care products: Concentrations and possible dermal exposure. *Environ. Res.* **2011**, *111*, 329–336. [CrossRef]
73. U.S. National Library of Medicine. National Center for Biotechnology Information. PubChem, 2019. Available online: https://pubchem.ncbi.nlm.nih.gov/ (accessed on 5 December 2019).
74. Luongo, G.; Östman, C. Organophosphate and phthalate esters in settled dust from apartment buildings in Stockholm. *Indoor Air* **2016**, *26*, 414–425. [CrossRef]
75. Brouwere, K.; Standaert, A.; Torfs, R. *Integrated Exposure for Risk Assessment in Indoor Environment (INTERA)*; Final Report; INTERA: Boeretang, Belgium, 2012.
76. Kolarik, B.; Bornehag, C.G.; Naydenov, K.; Sundell, J.; Stavova, P.; Nielsen, O.F. The concentrations of phthalates in settled dust in Bulgarian homes in relation to building characteristic and cleaning habits in the family. *Atmos. Environ.* **2008**, *42*, 8553–8559. [CrossRef]
77. Xu, Y.; Cohen Hubal, E.A.; Little, J.C. Predicting residential exposure to phthalate plasticizer emitted from vinyl flooring: Sensitivity, uncertainty, and implications for biomonitoring. *Environ. Health Perspect.* **2010**, *118*, 253–258. [CrossRef]
78. Weschler, C.J. Indoor/outdoor connections exemplified by processes that depend on an organic compound's saturation vapor pressure. In Proceedings of the Atmospheric Environment; Elsevier: Amsterdam, The Netherlands, 2003; Volume 37, pp. 5455–5465.
79. Institute for Health and Consumer Protection, Toxicology and Chemical Substances. Benzyl Butyl Phthalate. In *Summary Risk Assessment Report*; EUR 22773 EN/2; Institute for Health and Consumer Protection, Toxicology and Chemical Substances, European Chemicals Bureau: Ispra, Italy, 2008.
80. Santillo, D.; Labunska, I.; Davidson, H.; Johnston, P.; Strutt, M.; Knowles, O. *Consuming Chemicals–Hazardous Chemicals in House Dust as an Indicator of Chemical Exposure in the Home*; Greenpeace Research Laboratories: Exeter, UK, 2003.
81. European Chemicals Agency. *Evaluation of New Scientific Evidence Concerning DINP and DIDP in Relation to Entry 52 of Annex XVII to Reach Regulation (EC) No 1907/2006*; ECHA: Helsinki, Finland, 2013.
82. Fromme, H.; Lahrz, T.; Piloty, M.; Gebhart, H.; Oddoy, A.; Rüden, H. Occurrence of phthalates and musk fragrances in indoor air and dust from apartments and kindergartens in Berlin (Germany). *Indoor Air* **2004**, *14*, 188–195. [CrossRef] [PubMed]
83. Gevao, B.; Al-Ghadban, A.N.; Bahloul, M.; Uddin, S.; Zafar, J. Phthalates in indoor dust in Kuwait: Implications for non-dietary human exposure. *Indoor Air* **2013**, *23*, 126–133. [CrossRef] [PubMed]
84. Kanazawa, A.; Saito, I.; Araki, A.; Takeda, M.; Ma, M.; Saijo, Y.; Kishi, R. Association between indoor exposure to semi-volatile organic compounds and building-related symptoms among the occupants of residential dwellings. *Indoor Air* **2010**, *20*, 72–84. [CrossRef] [PubMed]
85. Yadav, I.C.; Devi, N.L.; Li, J.; Zhang, G. Polycyclic aromatic hydrocarbons in house dust and surface soil in major urban regions of Nepal: Implication on source apportionment and toxicological effect. *Sci. Total Environ.* **2018**, *616*, 223–235. [CrossRef]
86. Mannino, M.R.; Orecchio, S. Polycyclic aromatic hydrocarbons (PAHs) in indoor dust matter of Palermo (Italy) area: Extraction, GC-MS analysis, distribution and sources. *Atmos. Environ.* **2008**, *42*, 1801–1817. [CrossRef]

87. Maertens, R.M.; Yang, X.; Zhu, J.; Gagne, R.W.; Douglas, G.R.; White, P.A. Mutagenic and carcinogenic hazards of settled house dust I: Polycyclic aromatic hydrocarbon content and excess lifetime cancer risk from preschool exposure. *Environ. Sci. Technol.* **2008**, *42*, 1747–1753. [CrossRef]
88. Qi, H.; Li, W.L.; Zhu, N.Z.; Ma, W.L.; Liu, L.Y.; Zhang, F.; Li, Y.F. Concentrations and sources of polycyclic aromatic hydrocarbons in indoor dust in China. *Sci. Total Environ.* **2014**, *491*, 100–107. [CrossRef]
89. Vicente, E.D.; Alves, C.A. An overview of particulate emissions from residential biomass combustion. *Atmos. Res.* **2018**, *199*, 159–185. [CrossRef]
90. Cheruyiot, N.K.; Lee, W.J.; Mwangi, J.K.; Wang, L.C.; Lin, N.H.; Lin, Y.C.; Cao, J.; Zhang, R.; Chang-Chien, G.P. An overview: Polycyclic aromatic hydrocarbon emissions from the stationary and mobile sources and in the ambient air. *Aerosol Air Qual. Res.* **2015**, *15*, 2730–2762. [CrossRef]
91. Ong, S.; Ayoko, G.; Kokot, S.; Morawska, L. Polycyclic aromatic hydrocarbons in house dust samples: Source identification and apportionment. In Proceedings of the 14th International IUAPPA World Congress, Brisbane, QLD, Australia, 9–13 September 2007.
92. Ali, N.; Ismail, I.M.I.; Khoder, M.; Shamy, M.; Alghamdi, M.; Costa, M.; Ali, L.N.; Wang, W.; Eqani, S.A.M.A.S. Polycyclic aromatic hydrocarbons (PAHs) in indoor dust samples from Cities of Jeddah and Kuwait: Levels, sources and non-dietary human exposure. *Sci. Total Environ.* **2016**, *573*, 1607–1614. [CrossRef]
93. Qu, C.; Qi, S.; Yang, D.; Huang, H.; Zhang, J.; Chena, W.; Yohannes, H.K.; Sandy, E.H.; Yang, J.; Xing, X. Risk assessment and influence factors of organochlorine pesticides (OCPs) in agricultural soils of the hill region: A case study from Ningde, southeast China. *J. Geochem. Explor.* **2015**, *149*, 43–51. [CrossRef]
94. Roy, D.; Seo, Y.C.; Sinha, S.; Bhattacharya, A.; Singh, G.; Biswas, P.K. Human health risk exposure with respect to particulate-bound polycyclic aromatic hydrocarbons at mine fire-affected coal mining complex. *Environ. Sci. Pollut. Res.* **2019**, *26*, 19119–19135. [CrossRef] [PubMed]

Article

Chemical Characteristics of $PM_{2.5}$ and Water-Soluble Organic Nitrogen in Yangzhou, China

Yuntao Chen [1], Yanfang Chen [1], Xinchun Xie [1], Zhaolian Ye [2], Qing Li [2], Xinlei Ge [1,*] and Mindong Chen [1]

1 Jiangsu Key Laboratory of Atmospheric Environment Monitoring and Pollution Control, Collaborative Innovation Center of Atmospheric Environment and Equipment Technology, School of Environmental Sciences and Engineering, Nanjing University of Information Science and Technology, Nanjing 210044, China; chenyuntaonuist@163.com (Y.C.); chenyf920121@163.com (Y.C.); 13776502868@163.com (X.X.); chenmd@nuist.edu.cn (M.C.)

2 College of Chemistry and Environmental Engineering, Jiangsu University of Technology, Changzhou 213001, China; bess_ye@jsut.edu.cn (Z.Y.); qingli625@163.com (Q.L.)

* Correspondence: caxinra@163.com; Tel.: +86-25-58731394

Received: 7 March 2019; Accepted: 29 March 2019; Published: 3 April 2019

Abstract: Chemical characterization of fine atmospheric particles ($PM_{2.5}$) is important for effective reduction of air pollution. This work analyzed $PM_{2.5}$ samples collected in Yangzhou, China, during 2016. Ionic species, organic matter (OM), elemental carbon (EC), and trace metals were determined, and an Aerodyne soot-particle aerosol mass spectrometer (SP-AMS) was introduced to determine the OM mass, rather than only organic carbon mass. We found that inorganic ionic species was dominant (~52%), organics occupied about 1/4, while trace metals (~1%) and EC (~2.1%) contributed insignificantly to the total $PM_{2.5}$ mass. Water-soluble OM appeared to link closely with secondary OM, while water-insoluble OM correlated well with primary OM. The $PM_{2.5}$ concentrations were relatively low during summertime, while its compositions varied little among different months. Seasonal variations of water-soluble organic nitrogen (WSON) concentrations were not significant, while the mass contributions of WSON to total nitrogen were remarkably high during summer and autumn. WSON was found to associate better with secondary sources based on both correlation analyses and principle component analyses. Analyses of potential source contributions to WSON showed that regional emissions were dominant during autumn and winter, while the ocean became relatively important during spring and summer.

Keywords: $PM_{2.5}$; organic aerosol; water-soluble ions; organic nitrogen; aerosol mass spectrometry

1. Introduction

In recent years, particular matter (PM) pollution has gained wide attention from both government and the public due to its impacts on air quality, human health, and climate change [1–3]. In particular, much effort has been devoted to the study of fine particles with an aerodynamic diameter less than 2.5 μm ($PM_{2.5}$) due to its adverse health effects. Along with rapid economic growth, $PM_{2.5}$ pollution events ('haze') have occurred frequently in China, and many studies have been carried out to elucidate the properties of $PM_{2.5}$, especially in densely populated regions [4–9]. However, the chemical composition of $PM_{2.5}$ is highly complex, possibly containing a wide variety of components, including inorganic salts (nitrate, sulfate, and ammonium), trace metals, elemental carbon, and a large number of organic species. Moreover, the relative contributions of different components may change significantly under different atmospheric environments [10,11]. Local $PM_{2.5}$ properties have been shown to be governed by many factors, such as meteorological conditions, emission sources, boundary layer heights, and atmospheric transportation [12–14]. Therefore, comprehensive chemical analysis of $PM_{2.5}$, with a focus on particular geographic locations is necessary and essential for a better understanding of aerosol chemistry and its impacts.

Among various $PM_{2.5}$ components, the water-soluble organic nitrogen (WSON) species is an important group of compounds [15–17]. The mass fraction of WSON can account for ~10% to 68% of total aerosol nitrogen from continental to oceanic sites, highlighting the importance of organic nitrogen (ON) in atmospheric aerosols [18–21]. Some certain ON species, for example, amino compounds [22], have been shown to be very important for atmospheric new particle formation [23], secondary organic aerosol formation [24], aerosol hygroscopicity [25,26], and toxicity [27]. However, investigations on ON characteristics in China are scarce [28]. To the best of our knowledge, only one study, conducted in Changzhou [29], has reported on the WSON concentrations in the Yangtze River Delta (YRD) region.

In this study, we conducted filter sampling and chemical characterization of the $PM_{2.5}$ samples collected in urban Yangzhou during 2016. Yangzhou is located in the northern YRD region and is a major city with a population of ~4.6 million and an area of 6634 km^2. The city is a few hundred kilometers from Shanghai, and although it is a famous tourist city, it has also suffered recently from serious air pollution. Moreover, as an inland city, its air quality is influenced by multiple sources, including the transportation of air pollutants from nearby regions. However, even with these ongoing issues and concerns, chemical characterizations of the fine particles that are present in Yangzhou are quite scarce. In this paper, we applied a suite of analytical tools to achieve a relatively comprehensive analysis of the $PM_{2.5}$ components. Our analysis included a study of the following: Organic matter, elemental carbon, inorganic ions, and trace metals. In particular, we determined the mass concentration of the WSON. As part of our discussion, we highlight our analysis of the concentration of the WSON, with a particular focus on its seasonal variations, sources, and potential source areas. Our findings are a valuable contribution to the effort to reduce air pollution in Yangzhou and advance our understanding of aerosol properties in eastern China.

2. Experiments

2.1. Sampling Site and $PM_{2.5}$ Collection

The sampling site is located in urban Yangzhou (119.40° N, 32.38° E), next to the Yangtze Road (~300 m) and the Slender West Lake (~1 km) (Figure 1). The site is surrounded by residential areas, not directly influenced by industrial emissions. Samples for $PM_{2.5}$ analyses were collected from 6 April 2016 to 2 November 2016 by using a high volume (flow rate of 1.05 m^3 min^{-1}) sampler (Jinshida Ltd., Qingdao, China, model KB-1000) from 9.00 a.m. to 7.00 a.m. of the next day. This timeframe ensured a duration for each sample of 22 h. All particles were collected on 8 × 10-inch quartz fiber filters (Pallflex, Ann Arbor, MI, USA), which were prebaked at 450 °C in a muffle furnace for 4 h. A total of 128 samples were collected (no samples on precipitation days). Note, $PM_{2.5}$ samples during 13 November 2015–5 April 2016 were analyzed previously [30], and the results of that study were directly used in our study for comparison. For WSON analyses, samples across a whole year (232 samples) were used.

The $PM_{2.5}$ mass concentrations were determined gravimetrically by using a digital balance (OHAUS DV215CD, precision 0.01 mg) at 45% relative humidity (RH) and room temperature. Two filed blanks were treated in the same way as for the samples. All filters were wrapped in aluminum foil and stored in a freezer at −18 °C until analyzed. Hourly meteorological parameters were also recorded at the same site.

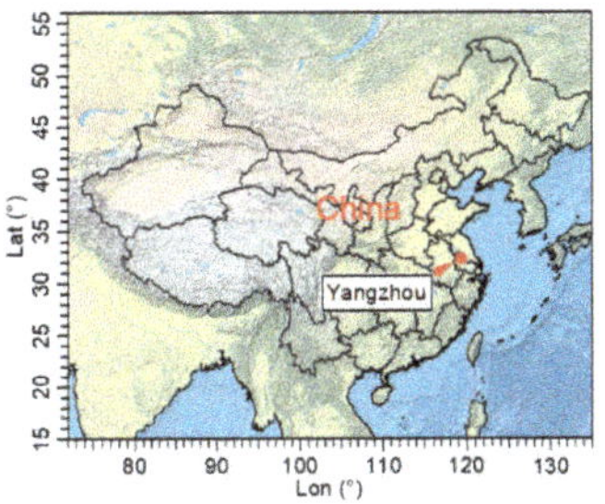

Figure 1. Sampling site and its surrounding areas.

2.2. Chemical Analyses

Ionic species. 1/4 filter was extracted with 100 mL ultrapure water (18.2 MΩ cm^{-1}) for 45 min ultrasonication in an ice-water bath. The solution was then filtered through a 0.45 μm syringe filter (Spartan, Whatman). An ion chromatograph (IC, Dionex ICS-600 for anions and ICS-1500 for cations) was used to measure the concentrations of water-soluble inorganic ions (WSIIs, including Na^+, NH_4^+, K^+, Mg^{2+}, Ca^{2+}, and F^-, Cl^-, NO_2^-, NO_3^-, SO_4^{2-}) in the aqueous extract. The operational details and method detection limits of these ionic species are the same as those described in Ye et al. [29].

Water-soluble organic carbon (WSOC) and total nitrogen (WSTN). A quarter of each filter was pre-treated in the same way as for ionic species. Following this, the WSOC concentrations were determined by a TOC-VCPH analyzer (Shimazu, Japan) using a thermo-catalytic oxidation approach. TN contents were determined by the TNM-1 unit of the analyzer. Instrument details and analysis procedures can be found in Ge et al. [31].

Organic carbon (OC) and elemental carbon (EC). A round piece with a diameter of 17 mm was cut from each filter and analyzed for total OC and EC contents by the thermal-optical OC/EC analyzer (Sunset Laboratory, Hillsborough, NC, USA) following the IMPROVE TOR protocol described in Chow et al. [32].

Trace elements. To measure the concentrations of trace metals, a MARS6 Xpress CEM corporation Microwave Digestion System was used to digest the $PM_{2.5}$ samples. About 2 cm^2 of each filter was cut into pieces and put into the polytetrafluoroethylene (PTFE) pressure digestion tank with 1 mL 49% hydrofluoric acid (HF) and 5 mL 69% nitric acid (HNO_3). The digest tank processed 8 samples in a batch. A Thermo Fisher X2 Series Inductively Coupled Plasma Mass Spectrometry (ICP-MS) was then used to determine the concentrations of 14 trace elements, including Al, V, Cr, Mn, Co, Ni, Cu, Zn, As, Se, Sr, Cd, Ba, and Pb. More details and detection limits of the elements can be found in Qi et al. [33].

Water-soluble organic matter (WSOM). Moreover, we introduced an Aerodyne soot-particle aerosol mass spectrometer [34] to obtain the composition of WSOM. Recently, this technique has been used for $PM_{2.5}$ samples in China [29,30,35–38]. Briefly, the water extracts were nebulized with argon by using a constant output atomizer (TSI Model 3076) and then dehumidified by a diffusion dryer filled with silica-gel. The dried particles were sent to the soot particle aerosol mass spectrometer (SP-AMS) for analysis. This instrument can output 70 ev electron impact ionization mass spectrum, and elemental ratios of the WSOM can be calculated based on the mass spectrum. More operational details can be found in our previous work [29,30].

2.3. Data Analyses

WSOM and water-insoluble OM (WIOM) estimations. The SP-AMS data were processed with the Igor-based ToF-AMS Analysis Toolkit (SQUIRREL v.1.57A and Pika v.1.16A) [39]. CO^+ was directly determined as we used pure argon as a carrier gas. Therefore, the CO^+ signal was not influenced by the N_2^+ signal and was easily quantified. For the organic CO_2^+ signal, since it can be influenced by carbonates [40] and nitrate [41], we set it equal to CO^+, similar to Aiken et al. [42]. The signals of H_2O^+, HO^+ and O^+ were scaled to CO_2^+ according to Aiken et al. [42]: $H_2O^+ = 0.225 \times CO_2^+$, $HO^+ = 0.05625 \times CO_2^+$, and $O^+ = 0.009 \times CO_2^+$. After determining WSOM mass spectrum, the oxygen-to-carbon (O/C) and hydrogen-to-carbon (H/C) ratios were calculated based on the method of Canagaratna et al. [43], while nitrogen-to-carbon (N/C) ratio was calculated based on Aiken et al. [42]. The organic mass-to-organic carbon (OM/OC) ratio was then calculated. Therefore, in a different approach from the traditional method, which often assumes a fixed OM/OC ratio (for example, 1.6 [44]), we are able to determine the OM/OC ratio for each sample by using the SP-AMS. However, it should be noted that these estimates were still empirical and are subject to uncertainties.

The SP-AMS directly-measured WSOM concentrations were affected by the concentrations of aqueous extracts and flow rates of the carrier gas, etc., Therefore, they cannot represent the original WSOM ambient mass loadings. As a result of this, we used the OM/OC ratios and TOC concentrations to calculate the WSOM concentrations:

$$WSOM = WSOC \times OM/OC_{WSOM} \tag{1}$$

The WIOM concentrations can be determined by the following equation:

$$WIOM = (OC - WSOC) \times 1.3 \tag{2}$$

The difference between total OC determined by the OC/EC analyzer and the WSOC, is water-insoluble OC (WIOC). The factor of 1.3 is used to convert WIOC to WIOM [45].

Primary (POM) and secondary OM (SOM) estimations. The OC and EC concentrations measured by the OC/EC analyzer can be used to infer the primary and secondary origins of OC by the EC-tracer method [46], as follows:

$$POC = EC \times (OC/EC)_{pri} \tag{3}$$

$$SOC = OC - POC \tag{4}$$

The $(OC/EC)_{pri}$ in Equation (3) refers to the value of primary OC (POC), which is typically the minimum value among all samples (1.92 in this study). The OC/EC ratios can vary among different sources and some primary organic aerosols, such as those from coal combustion and biomass burning may also have large OC/EC ratios. However, the uncertainty of POC and SOC estimations should be small according to Wu and Yu [47]. We can further estimate the POM and SOM concentrations [30]:

$$POM = POC \times 1.2 \tag{5}$$

$$SOM = (WSOM + WIOM) - POM \tag{6}$$

WSON estimations. The WSON concentrations can be calculated by the following equation:

$$WSON = WSTN - WSIN \tag{7}$$

In this study, concentrations of WSTN came from the TOC analyzer, while the concentrations of water-soluble inorganic nitrogen (WSIN) were the mass concentrations of nitrogen from NH_4^+, NO_3^-, and NO_2^- measured by the IC. Negative values (20 out of 232 samples) were not included.

Air mass trajectory and potential source contribution analyses of WSON. 72 h back trajectories of the air masses (at 500 m height) were obtained by using the Hybrid Single-particle Lagrangian Integrated Trajectory (HYSPLIT) model [48], based on the meteorological data downloaded from the National Oceanic and Atmospheric Administration (NOAA) Global Data Assimilation System (GDAS). The potential source contribution function (PSCF) analysis was performed to explore the air mass origins, and to identify potential source areas of WSON. Details of the PSCF analyses are described in Ge et al. [30].

3. Results and Discussion

3.1. General Characteristics of $PM_{2.5}$ Components

In our previous work [30], we focused on the characteristics of $PM_{2.5}$ in relatively polluted periods of autumn and winter; while this work presents results of less polluted periods, mainly spring and summer. Therefore, we were able to compare the $PM_{2.5}$ chemical characteristics between polluted and less polluted periods in a one-year span in Yangzhou. During the sampling period, wind speeds varied from 0.2 m/s to 5.2 m/s with an average of 1.6 m/s. RH had a mean value of 70%, which was on average higher than 62% during autumn–winter [30]. Average temperature was 24.2 °C (0 to 39.9 °C), being much higher than the 9.2 °C during autumn–winter [30]. The higher RH and temperature can influence the aerosol formation and gas-particle partitioning. Therefore, the aerosol properties could be different. Average precipitation was 2.63 mm, more than that during November 2015 to April 2016 [30], and most of the rainfall events occurred in June and July.

Figure 2a shows the time series of each major $PM_{2.5}$ component, and the reconstructed and weighted $PM_{2.5}$ mass concentrations. The average $PM_{2.5}$ concentration was 60.9 (±32.0) μg/m^3, ranging from 10.1 to 178.5 μg/m^3. This concentration was much lower than the mean value of 104 μg/m^{-3} during autumn–winter [30]. This result can be expected air pollution during autumn–winter is typically heavy [11] due to unfavorable meteorological conditions and increased primary emissions, such as coal combustion, etc. The summed mass concentrations of all measured components, including OM (=WSOM + WIOM), EC, ionic species and trace metals, were found to be correlated well with the gravimetrically determined $PM_{2.5}$ concentrations (Pearson's r = 0.94) (Figure 2b). However, the slope was only 0.80 (±0.02), indicating ~20% $PM_{2.5}$ mass was unidentified. This unknown fraction may include some unmeasured species, such as crustal materials, carbonate, and particle-bonded water, etc. In particular, organosulfates, recently recognized as a possible important component in $PM_{2.5}$ [49–51], may also contribute to this unknown portion. Sampling loss, analysis artifacts, and measurement uncertainties of various instruments may also add to this unknown portion. A previous similar study [38] also included an unidentified portion, sometimes up to 35.5% (results for $PM_{2.5}$ during January 2013 in Xi'an, China).

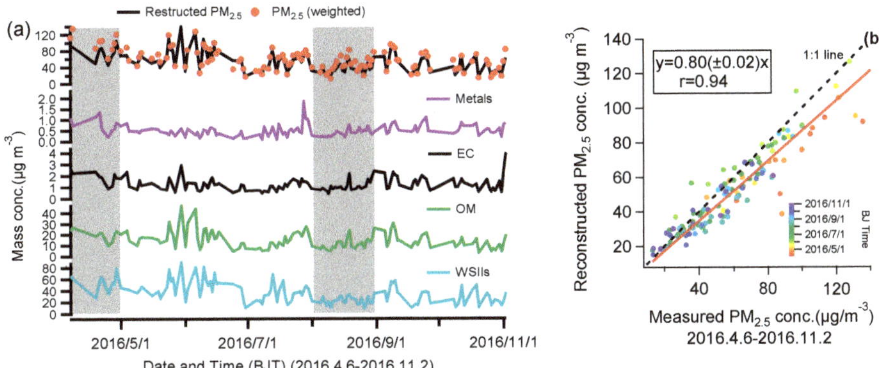

Figure 2. (**a**) Time series of total particular matter ($PM_{2.5}$) and major components, and (**b**) scatter plot of the reconstructed versus measured $PM_{2.5}$ mass concentrations (colored by time).

As shown in Figure 3, on average, the WSIIs (=Na^+ + $NH_4{}^+$ + K^+ + Mg^{2+} + Ca^{2+} + F^- + Cl^- + $NO_2{}^-$ + $NO_3{}^-$ + $SO_4{}^{2-}$) occupied the largest proportion (51.8%) of $PM_{2.5}$ mass; OM took up 24.6%, EC occupied 2.1%, and trace metals occupied only ~1%. The structure of $PM_{2.5}$ composition, and the relative abundances of these components, were similar to those observed during the autumn–winter period [30], revealing the general and consistent behavior of $PM_{2.5}$ in Yangzhou. Among WSIIs, sulfate was the most abundant component (36.4% of WSIIs), followed by nitrate (27.6%), and ammonium

(21.4%). The sum of sulfate, nitrate, and ammonium (SNA) was 85.4% of WSIIs mass, which was similar to 80.6% for the WSIIs during autumn–winter [30]. However, on the contrary, nitrate was more abundant than sulfate during autumn–winter. This result is consistent with previous findings in Changzhou [29] and other cities [52], since ammonium nitrate tends to evaporate into a gas phase at warm temperatures. Chloride fraction (3.2%) was also lower than that of autumn–winter [30], likely due to enhanced primary emissions of chloride, such as coal combustion. In Figure 4, we further presented the monthly averaged $PM_{2.5}$ chemical compositions during the sampling period. First, it can be seen that the $PM_{2.5}$ concentrations were low during summer (June–August). The relative contributions of different components varied, especially during June, the mass portion of WSIIs occupied ~70% $PM_{2.5}$ mass. EC and metal contributions were a bit larger during September and October than in other months. However, the average chemical composition of the most polluted month (April) was almost the same as that of the cleanest month (August), again indicating the consistent $PM_{2.5}$ characteristics in Yangzhou. It is important to note that daily samples were unable to capture the fast-chemical changes during pollution episodes. These samples are more likely to be representative of the average aerosol properties. Detailed characterization of aerosol properties during specific episodes can be investigated by highly time-resolved measurements, such as online AMS measurements [53–55].

Besides SNA, another relatively rich ionic species was calcium, which took up ~8.2% WSIIs mass, followed by minor contributions from Mg^{2+} (0.3%), K^+ (1.0%), and Na^+ (1.2%) (Figure 3). Other trace metals together, only occupied ~1% of total $PM_{2.5}$ mass, which was lower than those reported in Changzhou (2.7% and 5.0%) [29], etc., but similar to that found in Nanjing (1%) [33]. Zn, Al, Pb, and Mn were relatively rich (together 75.4%), and the other 10 elements only occupied ~1/4 mass of the trace elements. It is important to note that during autumn–winter [30], Zn, Al, Pb, and Mn were also the top four abundant trace elements, but the relative contributions were very different, indicating different sources at different seasons for the trace metals in Yangzhou.

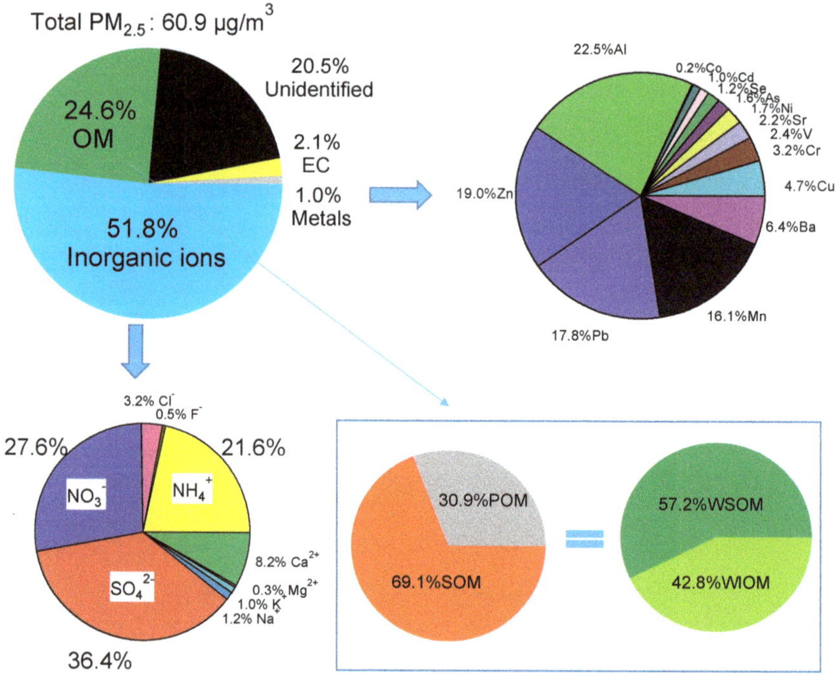

Figure 3. Average chemical compositions of total $PM_{2.5}$, organic matter, ionic species, and trace metals (2016.4.6~2016.11.2).

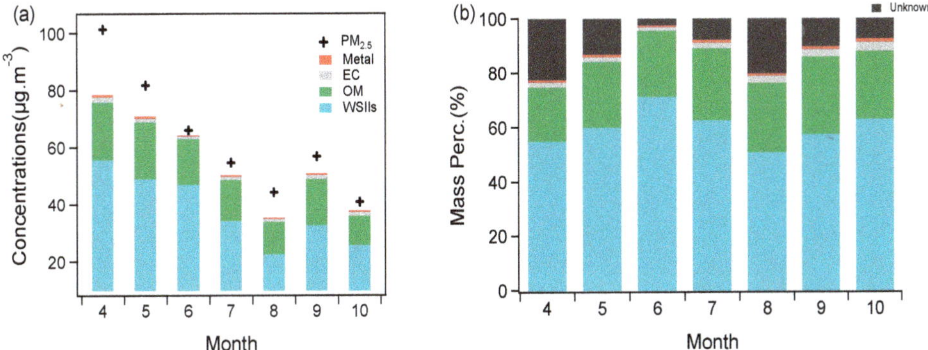

Figure 4. Monthly averaged chemical compositions and mass concentrations of $PM_{2.5}$ (**a**), and mass contributions of different components to $PM_{2.5}$ (**b**).

By using the OC/EC-tracer based method and the SP-AMS measured results, we can separate OM mass into POM and SOM, as well as WSOM and WIOM. SOM dominated over POM, and WSOM dominated the OM (Figure 3), both were similar to the results during autumn–winter [30]. Tight correlations were observed between WSOM and SOM (r of 0.96 and slope of 0.81) (Figure 5a), and between WIOM and POM (r of 0.80 and slope of 1.37) (Figure 5b). A possible reason for this, is that the process generally involved in the generation of SOM from POM is oxidation, which increases polarity (e.g., generation of carboxylates). The parent compounds (POM) might generally be expected to be non-polar (e.g., polycyclic aromatic compounds and long-chain hydrocarbons) and so would be expected to correlate well with WIOM. This result links POM and SOM with their hygroscopicities, and demonstrates that SOM is likely hygroscopic, while POM is probably hydrophobic. Of course, higher SOM fraction than WSOM fraction (69.1% vs. 57.2%) indicates that a portion of SOM can be water-insoluble.

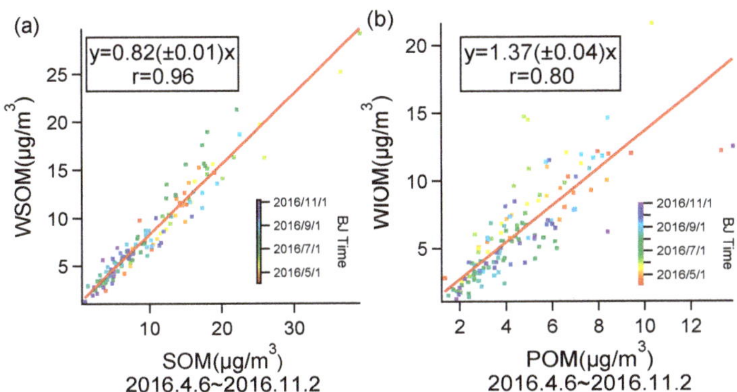

Figure 5. Scatter plots of water-soluble organic matter (WSOM) versus secondary OM (SOM) (**a**), and water-insoluble OM (WIOM) versus primary OM (POM) (**b**) (colored by time).

In ambient air, ammonia is typically the major basic compound to neutralize sulfuric, nitric, and hydrochloric acids. In this work, the equivalent molar ratios of ammonium to the sum of sulfate, nitrate and chloride, was on average 0.95 and the correlation was also good (r of 0.92) (Figure 6a). This result manifests that sulfate, nitrate, and chloride are dominantly bonded with ammonium. The fitted slope of 0.95 (±0.03), less than 1, likely indicates a slight deficiency of ammonium, and suggests a small portion of sulfate, etc., may bond with other cations, such as Ca^{2+}, according to Figure 3.

Considering all measured ionic species, the ionic balance can be calculated by using the molar equivalent ratios of cations (CE) to anions (AE), as follows:

$$\mathrm{CE} = \frac{[\mathrm{Na^+}]}{23} + \frac{[\mathrm{NH_4}^+]}{18} + \frac{[\mathrm{K^+}]}{39} + \frac{[\mathrm{Ca^{2+}}]}{40} \times 2 + \frac{[\mathrm{Mg^{2+}}]}{24} \times 2 \quad (8)$$

$$\mathrm{AE} = \frac{[\mathrm{SO_4}^{2-}]}{96} \times 2 + \frac{[\mathrm{NO_3}^-]}{62} + \frac{[\mathrm{Cl^-}]}{35.5} + \frac{[\mathrm{F^-}]}{19} \quad (9)$$

It is important to note that NO_2^- concentration was below the detection limit and was, therefore, not included in Equation (9). The scatter plot between CE and AE is presented in Figure 6b. The correlation was very tight (r of 0.95), and the slope of was 1.06 (±0.03), a bit larger than 1. The result shows that measured cations are able to neutralize all measured anions, and also suggests the possible existence of some other undetected anions (probably carbonate and organic anions).

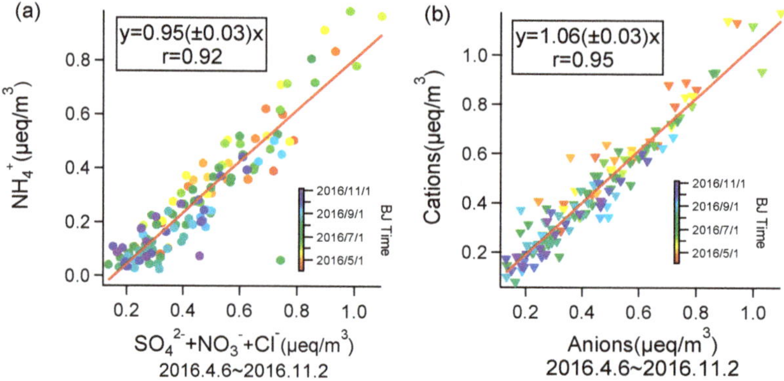

Figure 6. Scatter plots of molar ratios of ammonium versus sum of sulfate, nitrate, and chloride (**a**), and cations versus anions (**b**) (colored by time).

3.2. Seasonal Properties and Sources of WSON

By using the method provided in Section 2.3, we calculated the WSON concentrations for all samples across a full year. The annual mean WSON concentration was determined to 1.71 (±1.08) μg/m^{-3}. Table 1 compares our results with a number of previous studies conducted in China and a few other locations. It can be seen that, the WSON level in Yangzhou was much higher than those in Japan, Greece, and Florida, USA. It was also higher than ON in PM_1 in Beijing, but this was likely due to the fact that ON in supermicronmeter aerosols was not included in that study. This argument is furthered by another study on $PM_{2.5}$ in Beijing, which showed a much higher WSON concentration. The WSON concentration in Yangzhou was also much lower than those in Xi'an and Guangzhou, but similar to that observed in Changzhou. However, the annual average mass percentage of WSON in WSTN was 18.7%, which is relatively high among the studies listed in Table 1 (except Beijing and Xi'an), indicating an important role of WSON in Yangzhou $PM_{2.5}$.

Table 1. Mass concentrations of water-soluble organic nitrogen (WSON) measured in different sites.

Location	Samples	Period	WSON ($\mu g/m^3$)	WSON/WSTN (%)	Ref.
Beijing, China	$PM_{2.5}$	1998.11–1999.2	3.17	30	[56]
Xi'an, China	$PM_{2.5}$	2008–2009	4.20	45	[21]
Guangzhou, China	$PM_{2.5}$	2014	4.78	12.2	[16]
Changzhou, China	$PM_{2.5}$	2015.7–2016.4	1.16	7.7	[29]
Beijing, China *	PM_1	2013.12.16–2014.12.15	0.26~0.59	7-10	[28]
East China Seas	TSP (Sea fog)	2014.3.17–2014.4.22	2.06 ± 2.39	10 ± 6	[57]
Rishiri Island, Japan	PM_{10}	2010.4.16–2012.11.8	0.077	13	[58]
Kofu, Japan	PM_{10}	2016.4.16–2016.10.13	0.071	10	[59]
Crete, Greece	$PM_{1.3}$	2005.1–2003.12	0.83 ± 1.0	13	[17]
Florida, USA	$PM_{2.5}$	2005.7–2005.9	0.36 ± 0.21	8.9 ± 5.8	[60]
Yangzhou, China	$PM_{2.5}$	2015.11.12–2016.11.2	1.71 ± 1.08	18.7	This work

* Online measurement of ON in PM_1. It may include both water-soluble and water-insoluble ON.

We present the seasonal average WSON concentrations in Figure 7a and the corresponding monthly averages in Figure 8a. The average concentrations were 1.87, 1.55, 1.66, and 1.70 $\mu g/m^{-3}$ in winter, spring, summer, and autumn, respectively. Winter had a slightly higher WSON concentration than in other seasons, but overall the seasonal or monthly differences of WSON concentrations were not significant except a higher concentration during December and a lower concentration during August. Furthermore, we illustrate the seasonal average nitrogen concentrations in ammonium (NH_4^+_N) and nitrate (NO_3^-_N) together with WSON in Figure 7b. In all seasons, NH_4^+_N was the most dominant N-containing species but displayed a decreasing trend from winter to autumn. NO_3^-_N concentration was second to NH_4^+_N and showed a similar trend with an exception in summer. During summer, WSON concentration was even higher than NO_3^-_N concentration, which is again attributed to the thermodynamic dissociation of NH_4NO_3 under summertime high temperatures. Overall, different from insignificant seasonal variations of WSON concentrations shown in Figure 7a, the mass contributions of WSON to WSTN were remarkably higher in summer (24.9%) and autumn (24.5%) than in winter (11.4%) and spring (12.6%). Monthly data, as can be seen in Figure 8b, further shows that the average mass percentages of WSON/WSTN were nearly 40% in August and September. This result underscores the importance of WSON in summer and autumn in Yangzhou, despite the total $PM_{2.5}$ concentrations were relatively low in these two seasons.

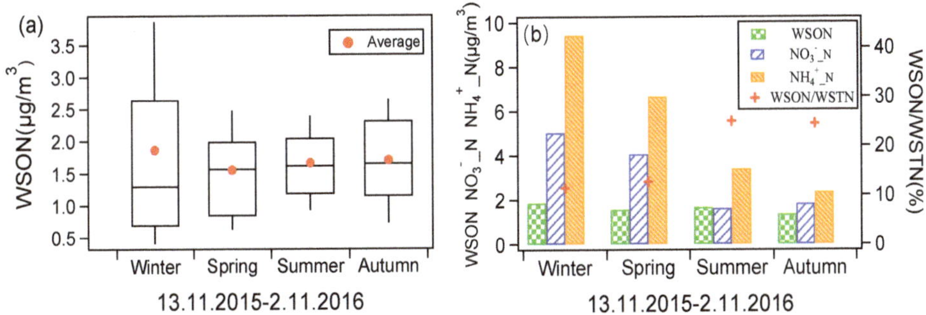

Figure 7. Seasonal average concentrations of WSON (**a**), and average concentrations of WSON, nitrogen in ammonium (NH_4^+_N) and nitrate (NO_3^-_N), and WSON/water-soluble total nitrogen (WSTN) (%) (**b**).

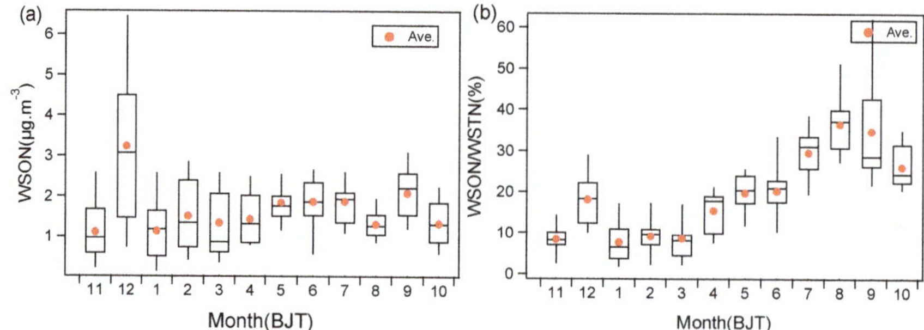

Figure 8. Monthly average concentrations of WSON (**a**), and WSON/WSTN (%) (**b**).

To explore the sources of WSON, we first investigated the correlations between WSON and SOM (also POM), as shown in Figure 9. The correlation of WSON with SOM (r of 0.49) was better than it was with POM (r of 0.26), indicating that WSON was more likely associated with secondary formation processes than with primary emissions in Yangzhou. However, the correlation coefficient (0.49) was only moderate, demonstrating that WSON sources were, in fact, complex. Therefore, we further used principle components analyses (PCA) to identify the possible sources of WSON (as well as $PM_{2.5}$). Twelve variables including OC, EC, WSOC, WSON, and major ionic species of Cl^-, NO_3^-, SO_4^{2-}, Na^+, NH_4^+, K^+, Mg^{2+}, and Ca^{2+} were chosen as input parameters.

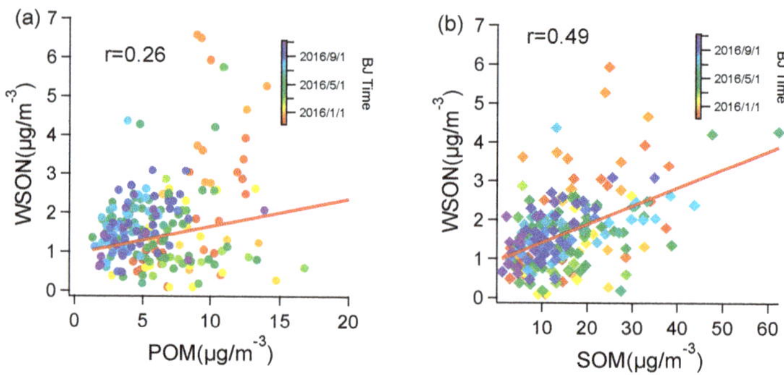

Figure 9. Scatter plots of WSON versus POM (**a**), and WSON versus SOM (**b**) (colored by time).

The PCA resolved four principle factors (or sources) with eigenvalues larger than 1, together accounting for 82.41% of the total variance of original data matrix. The factor profiles, representing the correlation coefficients between the input variables with the factor, are given in Table 2. Component 1 (PC1), which explained 53.1% of the variance, was very likely from biomass and fossil fuel combustions, as it associated closely with K^+ (biomass burning marker ion), OC, EC, and Cl^- (possible species from fossil fuel and or coal combustion). PC2 explained 11.77% of the total variance and the species with high loadings are SNA (SO_4^{2-}, NO_3^-, NH_4^+) and WSON. SNA are known to be generated from secondary reactions, therefore PC2 represents the secondary source. The third component (PC3, 9.34% variance) had high loadings of Mg^{2+} and Ca^{2+}, likely indicating dust origin (from soil and construction activities, etc.). For the fourth component (PC4, 8.21% variance), only Na^+ had a high loading (0.97), suggesting a sea salt source. Generally, for WSON, obviously higher loading in the secondary source (PC2) than in other primary sources was observed, strongly indicating that WSON was mainly from a secondary source, consistent with the correlation analysis results in Figure 8. Nevertheless, further molecular characterizations and quantifications of the WSON species and other source markers, combined with source apportionment techniques are still needed to provide more insights into the sources of WSON.

Table 2. Principle components analyses (PCA) analysis results of $PM_{2.5}$ and WSON (numbers in bold highlight the components with significances).

Components	PC1	PC2	PC3	PC4
EC	**0.94**	0.02	0.11	0.09
OC	**0.79**	0.36	0.21	−0.02
Cl^-	**0.88**	0.03	0.10	0.12
NO_3^-	**0.79**	**0.50**	−0.03	−0.11
SO_4^{2-}	0.42	**0.78**	0.02	−0.07
NH_4^+	**0.79**	**0.51**	−0.07	−0.04
Ca^{2+}	0.29	−0.02	**0.74**	0.17
Mg^{2+}	−0.03	0.06	**0.90**	−0.18
Na^+	0.04	0.06	−0.03	**0.97**
K^+	**0.93**	0.15	0.14	0.03
WSOC	**0.82**	0.42	0.11	−0.01
WSON	0.09	**0.79**	0.03	0.13
Initial Eigenvalue	6.37	1.41	1.12	1.00
Variance Explained	53.10	11.77	9.34	8.21
Cumulative Variance Explained	53.10	64.86	74.20	82.41

The potential source areas (PSCs) of WSON and clusters of air mass trajectories in four seasons are displayed in Figures 10 and 11, respectively. Five clusters were identified during winter, while other seasons only had three clusters. During winter, the WSON PSCs were mainly distributed within the YRD region (Figure 10a), indicating the dominance of regional emissions. But a few hotspots were also found in Shandong province and, correspondingly, two clusters, one originating from Beijing (42.86% of total trajectories) and one initiated from Mongolia (25.40% of total trajectories), both travelled across Shandong Province. Although the WSON PSCs during winter could trace back far beyond the YRD region and Shandong province, other areas contributed relatively insignificantly (correspondingly clusters 2, 4, and 5 together occupied ~30% of all trajectories in Figure 11a). No obvious source from the ocean was found during winter.

For spring, the WSON PSCs mainly distributed in the YRD region and Jiangxi province, rather than in Shandong Province (Figure 10b). Interestingly, there was a potentially large source in the Yellow sea close to Korea, and consistently a relatively short cluster starting from Bohai sea, crossing over Shandong Peninsula and the Yellow Sea, occupied the majority of air mass (62.75%) (Figure 11b).

The WSON PSCs during summer concentrated mainly in Jiangsu and Zhejiang provinces. There were also some WSON likely from the ocean, including the Yellow Sea and the East China Sea (Figure 10c). The trajectories during summer were also unique: It had one cluster (41.3%), which originated from the east Pacific Ocean, and another one (19.57%) crossed over Bohai Sea and Yellow Sea. There was only one inland cluster (41.3%), which started from Guangdong Province (Figure 11c).

During autumn, the WSON PSCs were predominantly located in Jiangsu Province and Shanghai, with less contributions from Shandong Province and the Yellow Sea. Consistently, a short cluster across the Yellow Sea, Shanghai, and Jiangsu Province, was the major one (55.88%). The area of WSON potential sources during autumn appeared to be the smallest among all seasons.

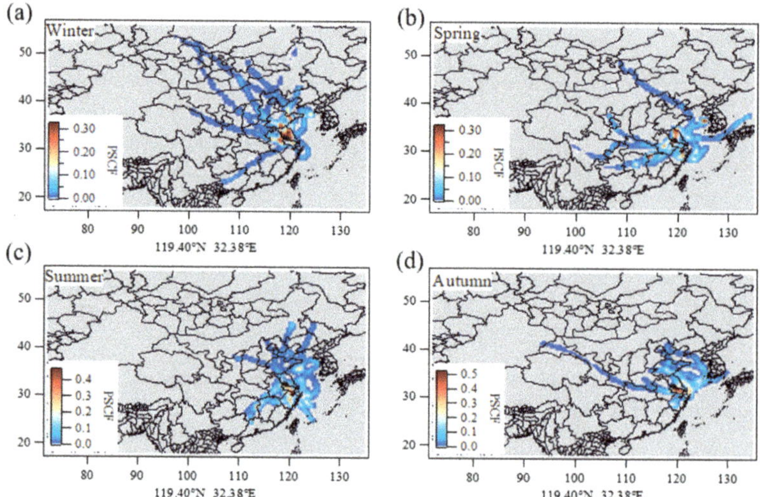

Figure 10. The potential source contributions to WSON during (**a**) winter, (**b**) spring, (**c**) summer, and (**d**) autumn (colored by the potential source contribution function (PSCF) values).

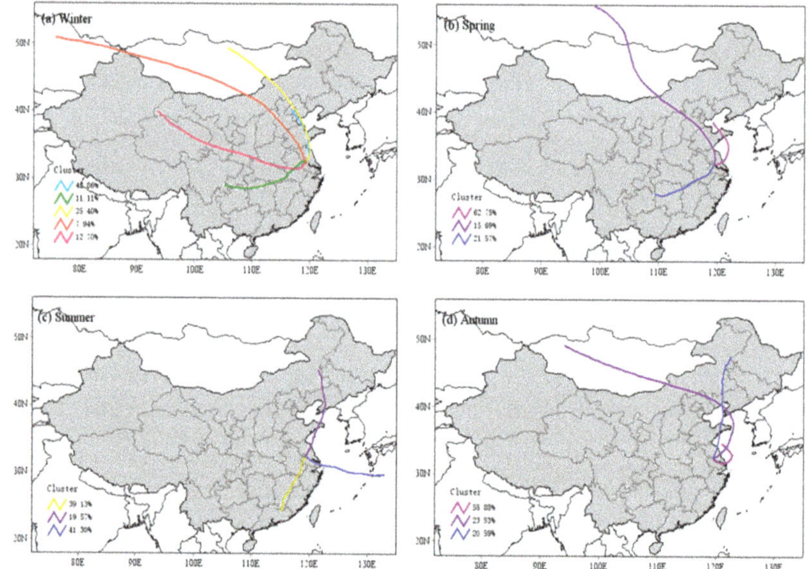

Figure 11. Clusters of the 72-h back trajectories during (**a**) winter, (**b**) spring, (**c**) summer, and (**d**) autumn.

4. Conclusions

In this work, we analyzed the chemical compostions of $PM_{2.5}$ samples collected in urban Yangzhou, China during spring and summer of 2016. Ionic species, OM, EC, and trace metals, on average. occupied ~80% of total $PM_{2.5}$ mass. Mass contribution of ionic species (~52%) was twice that of the OM (~25%), while EC and trace metals together contributed little (~3%). Sulfate, nitrate, and ammonium dominated the ionic species. The Aerodyne SP-AMS was used to determine the mass spectra of WSOM, and derived the OM/OC ratio, and was therefore the WSOM and WIOM mass concentrations. An EC-tracer based method was used to estimate SOM and POM mass concentrations. Correlation analyses showed that SOM linked closely with WSOM, while POM correlated well with WIOM. In particular, this work measured the WSON concentrations in one year-long $PM_{2.5}$ samples. Different from the monthly variations of $PM_{2.5}$ concentrations, WSON concentrations had no significant seasonal and monthly differences. However, seasonal and monthly differences of mass contributions of WSON to total nitrogen were significant, which were much higher in summer and autumn than in winter and spring. Temporal variation of WSON concentrations was found to correlate better with that of SOM than with POM, indicating secondary formation is important for the WSON in Yangzhou, which was also supported by the PCA results. PSCF analyses of WSON demonstrated that regional sources were dominant during autumn and winter, while oceans became relatively important for the WSON during spring and summer.

Author Contributions: Conceptualization, X.G. and M.C.; methodology, Y.C. (Yuntao Chen), Y.C. (Yanfang Chen), X.X., Q.L. and Z.Y.; sample collection and chemical analysis: Y.C. (Yuntao Chen), Y.C. (Yanfang Chen), X.X., Q.L. and Z.Y.; data analysis and discussion: Y.C. (Yuntao Chen), Y.C. (Yanfang Chen) and X.G.; writing—original draft preparation, Y.C. (Yuntao Chen); writing—review and editing, X.G. and M.C.; supervision, X.G.; project administration, X.G. and M.C.; funding acquisition, X.G.

Funding: This research was funded by the Natural Science Foundation of China (NSFC), grant number 21577065 and 21777073, and the Jiangsu Natural Science Foundation, grant number BK20181476.

Acknowledgments: The authors would like to thank the Environmental Monitoring Center of Yangzhou for the technical assistance provided for this work.

Conflicts of Interest: The authors declare no conflict of interest.

References

1. Heal, M.R.; Kumar, P.; Harrison, R.M. Particles, air quality, policy and health. *Chem. Soc. Rev.* **2012**, *41*, 6606–6630. [CrossRef] [PubMed]
2. Nel, A. Air pollution-related illness: Effects of particles. *Science* **2005**, *308*, 804–806. [CrossRef]
3. Cao, J.J.; Xu, H.M.; Xu, Q.; Chen, B.H.; Kan, H.D. Fine particulate matter constituents and cardiopulmonary mortality in a heavily polluted chinese city. *Environ. Health Persp.* **2012**, *120*, 373–378. [CrossRef] [PubMed]
4. Chan, C.K.; Yao, X. Air pollution in mega cities in china. *Atmos. Environ.* **2008**, *42*, 1–42. [CrossRef]
5. Liang, C.; Duan, F.; He, K.; Ma, Y. Review on recent progress in observations, source identifications and countermeasures of pm2.5. *Environ. Int.* **2016**, *86*, 150–170. [CrossRef] [PubMed]
6. Liu, X.; Nie, D.; Zhang, K.; Wang, Z.; Li, X.; Shi, Z.; Wang, Y.; Huang, L.; Chen, M.; Ge, X.; et al. Evaluation of particulate matter deposition in the human respiratory tract during winter in nanjing using size and chemically resolved ambient measurements. *Air Qual. Atmos. Health* **2019**. [CrossRef]
7. Nie, D.; Chen, M.; Wu, Y.; Ge, X.; Hu, J.; Zhang, K.; Ge, P. Characterization of fine particulate matter and associated health burden in nanjing. *Int. J. Environ. Res. Public Health* **2018**, *15*, 602. [CrossRef]
8. Sun, Y.; Du, W.; Fu, P.; Wang, Q.; Li, J.; Ge, X.; Zhang, Q.; Zhu, C.; Ren, L.; Xu, W.; et al. Primary and secondary aerosols in beijing in winter: Sources, variations and processes. *Atmos. Chem. Phys.* **2016**, *16*, 8309–8329. [CrossRef]
9. Chen, Y.; Xie, S.-D.; Luo, B.; Zhai, C. Characteristics and sources of water-soluble ions in pm2.5 in the sichuan basin, china. *Atmosphere* **2019**, *10*, 78. [CrossRef]
10. Saxena, P.; Hildemann, L.M. Water-soluble organics in atmospheric particles: A critical review of the literature and application of thermodynamics to identify candidate compounds. *J. Atmos. Chem.* **1996**, *24*, 57–109. [CrossRef]
11. Wang, Y.; Ying, Q.; Hu, J.; Zhang, H. Spatial and temporal variations of six criteria air pollutants in 31 provincial capital cities in china during 2013–2014. *Environ. Int.* **2014**, *73*, 413–422. [CrossRef] [PubMed]
12. Shen, F.; Ge, X.; Hu, J.; Nie, D.; Tian, L.; Chen, M. Air pollution characteristics and health risks in henan province, china. *Environ. Res.* **2017**, *156*, 625–634. [CrossRef] [PubMed]
13. Zhu, Y.; Huang, L.; Li, J.; Ying, Q.; Zhang, H.; Liu, X.; Liao, H.; Li, N.; Liu, Z.; Mao, Y.; et al. Sources of particulate matter in china: Insights from source apportionment studies published in 1987–2017. *Environ. Int.* **2018**, *115*, 343–357. [CrossRef]

14. Zhang, R.; Wang, G.; Guo, S.; Zamora, M.L.; Ying, Q.; Lin, Y.; Wang, W.; Hu, M.; Wang, Y. Formation of urban fine particulate matter. *Chem. Rev.* **2015**, *115*, 3803–3855. [CrossRef]
15. Cape, J.N.; Cornell, S.E.; Jickells, T.D.; Nemitz, E. Organic nitrogen in the atmosphere—Where does it come from? A review of sources and methods. *Atmos. Res.* **2011**, *102*, 30–48. [CrossRef]
16. Yu, X.; Yu, Q.; Zhu, M.; Tang, M.; Li, S.; Yang, W.; Zhang, Y.; Deng, W.; Li, G.; Yu, Y.; et al. Water soluble organic nitrogen (wson) in ambient fine particles over a megacity in south china: Spatiotemporal variations and source apportionment. *J. Geophys. Res. Atmos.* **2017**, *122*, 13,045–13,060. [CrossRef]
17. Violaki, K.; Mihalopoulos, N. Water-soluble organic nitrogen (wson) in size-segregated atmospheric particles over the eastern mediterranean. *Atmos. Environ.* **2010**, *44*, 4339–4345. [CrossRef]
18. Zhang, Q.; Anastasio, C.; Jimenez-Cruz, M. Water-soluble organic nitrogen in atmospheric fine particles (pm2.5) from northern california. *J. Geophys. Res. Atmos.* **2002**, *107*, 4112. [CrossRef]
19. Mace, K.A.; Duce, R.A.; Tindale, N.W. Organic nitrogen in rain and aerosol at cape grim, tasmania, australia. *J. Geophys. Res. Atmos.* **2003**, *108*, 4338. [CrossRef]
20. Rastogi, N.; Zhang, X.; Edgerton, E.S.; Ingall, E.; Weber, R.J. Filterable water-soluble organic nitrogen in fine particles over the southeastern USA during summer. *Atmos. Environ.* **2011**, *45*, 6040–6047. [CrossRef]
21. Ho, K.F.; Ho, S.S.H.; Huang, R.J.; Liu, S.X.; Cao, J.J.; Zhang, T.; Chuang, H.C.; Chan, C.S.; Hu, D.; Tian, L. Characteristics of water-soluble organic nitrogen in fine particulate matter in the continental area of china. *Atmos. Environ.* **2015**, *106*, 252–261. [CrossRef]
22. Ge, X.; Wexler, A.S.; Clegg, S.L. Atmospheric amines—Part i. A review. *Atmos. Environ.* **2011**, *45*, 524–546. [CrossRef]
23. Almeida, J.; Schobesberger, S.; Kurten, A.; Ortega, I.K.; Kupiainen-Maatta, O.; Praplan, A.P.; Adamov, A.; Amorim, A.; Bianchi, F.; Breitenlechner, M.; et al. Molecular understanding of sulphuric acid-amine particle nucleation in the atmosphere. *Nature* **2013**, *502*, 359–363. [CrossRef]
24. Qiu, C.; Zhang, R. Multiphase chemistry of atmospheric amines. *Phys.Chem. Chem. Phys.* **2013**, *15*, 5738–5752. [CrossRef] [PubMed]
25. Ge, X.; Wexler, A.S.; Clegg, S.L. Atmospheric amines—Part ii. Thermodynamic properties and gas/particle partitioning. *Atmos. Environ.* **2011**, *45*, 561–577. [CrossRef]
26. Fan, X.; Dawson, J.; Chen, M.; Qiu, C.; Khalizov, A. Thermal stability of particle-phase monoethanolamine salts. *Environ. Sci. Technol.* **2018**, *52*, 2409–2417. [CrossRef]
27. Lee, D.; Wexler, A.S. Atmospheric amines—Part iii: Photochemistry and toxicity. *Atmos. Environ.* **2013**, *71*, 95–103. [CrossRef]
28. Xu, W.; Sun, Y.; Wang, Q.; Du, W.; Zhao, J.; Ge, X.; Han, T.; Zhang, Y.; Zhou, W.; Li, J.; et al. Seasonal characterization of organic nitrogen in atmospheric aerosols using high resolution aerosol mass spectrometry in beijing, china. *ACS Earth Space Chem.* **2017**, *1*, 673–682. [CrossRef]
29. Ye, Z.; Liu, J.; Gu, A.; Feng, F.; Liu, Y.; Bi, C.; Xu, J.; Li, L.; Chen, H.; Chen, Y.; et al. Chemical characterization of fine particular matter in changzhou, china and source apportionment with offline aerosol mass spectrometry. *Atmos. Chem. Phys.* **2017**, *2017*, 2573–2592. [CrossRef]
30. Ge, X.; Li, L.; Chen, Y.; Chen, H.; Wu, D.; Wang, J.; Xie, X.; Ge, S.; Ye, Z.; Xu, J.; et al. Aerosol characteristics and sources in yangzhou, china resolved by offline aerosol mass spectrometry and other techniques. *Environ. Pollut.* **2017**, *225*, 74–85. [CrossRef]
31. Ge, X.; Shaw, S.L.; Zhang, Q. Toward understanding amines and their degradation products from postcombustion CO_2 capture processes with aerosol mass spectrometry. *Environ. Sci. Technol.* **2014**, *48*, 5066–5075. [CrossRef]
32. Chow, J.C.; Watson, J.G.; Chen, L.W.A.; Chang, M.C.O.; Robinson, N.F.; Trimble, D.; Kohl, S. The improve_a temperature protocol for thermal/optical carbon analysis: Maintaining consistency with a long-term database. *J. Air Waste Manag.* **2007**, *57*, 1014–1023. [CrossRef]
33. Qi, L.; Chen, M.; Ge, X.; Zhang, Y.; Guo, B. Seasonal variations and sources of 17 aerosol metal elements in suburban nanjing, china. *Atmosphere* **2016**, *7*, 153. [CrossRef]
34. Onasch, T.B.; Trimborn, A.; Fortner, E.C.; Jayne, J.T.; Kok, G.L.; Williams, L.R.; Davidovits, P.; Worsnop, D.R. Soot particle aerosol mass spectrometer: Development, validation, and initial application. *Aerosol. Sci. Technol.* **2012**, *46*, 804–817. [CrossRef]
35. Xu, J.; Zhang, Q.; Wang, Z.; Yu, G.; Ge, X.; Qin, X. Chemical composition and size distribution of summertime pm2.5 at a high altitude remote location in the northeast of the qinghai–xizang (tibet) plateau: Insights into aerosol sources and processing in free troposphere. *Atmos. Chem. Phys.* **2015**, *15*, 5069–5081. [CrossRef]

36. Qiu, Y.; Xie, Q.; Wang, J.; Xu, W.; Li, L.; Wang, Q.; Zhao, J.; Chen, Y.; Chen, Y.; Wu, Y.; et al. Vertical characterization and source apportionment of water-soluble organic aerosol with high-resolution aerosol mass spectrometry in beijing, china. *ACS Earth Space Chem.* **2019**, *3*, 273–284. [CrossRef]
37. Ye, Z.; Li, Q.; Liu, J.; Luo, S.; Zhou, Q.; Bi, C.; Ma, S.; Chen, Y.; Chen, H.; Li, L.; et al. Investigation of submicron aerosol characteristics in changzhou, china: Composition, source, and comparison with co-collected pm2.5. *Chemosphere* **2017**, *183*, 176–185. [CrossRef]
38. Huang, R.-J.; Zhang, Y.; Bozzetti, C.; Ho, K.-F.; Cao, J.-J.; Han, Y.; Daellenbach, K.R.; Slowik, J.G.; Platt, S.M.; Canonaco, F.; et al. High secondary aerosol contribution to particulate pollution during haze events in china. *Nature* **2014**, *514*, 218. [CrossRef] [PubMed]
39. Tof-ams Software. (Written in Igor Pro 6.37—Wavemetrics, Portland, OR, USA). Available online: http://cires1.Colorado.Edu/jimenez-group/tofamsresources/tofsoftware/index.Html (accessed on 9 January 2019).
40. Bozzetti, C.; El Haddad, I.; Salameh, D.; Daellenbach, K.R.; Fermo, P.; Gonzalez, R.; Minguillón, M.C.; Iinuma, Y.; Poulain, L.; Elser, M.; et al. Organic aerosol source apportionment by offline-ams over a full year in marseille. *Atmos. Chem. Phys.* **2017**, *17*, 8247–8268. [CrossRef]
41. Pieber, S.M.; El Haddad, I.; Slowik, J.G.; Canagaratna, M.R.; Jayne, J.T.; Platt, S.M.; Bozzetti, C.; Daellenbach, K.R.; Fröhlich, R.; Vlachou, A.; et al. Inorganic salt interference on $CO_2{}^+$ in aerodyne ams and acsm organic aerosol composition studies. *Environ. Sci. Technol.* **2016**, *50*, 10494–10503. [CrossRef]
42. Aiken, A.C.; Decarlo, P.F.; Kroll, J.H.; Worsnop, D.R.; Huffman, J.A.; Docherty, K.S.; Ulbrich, I.M.; Mohr, C.; Kimmel, J.R.; Sueper, D.; et al. O/c and om/oc ratios of primary, secondary, and ambient organic aerosols with high-resolution time-of-flight aerosol mass spectrometry. *Environ. Sci. Technol.* **2008**, *42*, 4478–4485. [CrossRef]
43. Canagaratna, M.R.; Jimenez, J.L.; Kroll, J.H.; Chen, Q.; Kessler, S.H.; Massoli, P.; Hildebrandt Ruiz, L.; Fortner, E.; Williams, L.R.; Wilson, K.R.; et al. Elemental ratio measurements of organic compounds using aerosol mass spectrometry: Characterization, improved calibration, and implications. *Atmos. Chem. Phys.* **2015**, *15*, 253–272. [CrossRef]
44. Ye, Z.; Li, Q.; Ma, S.; Zhou, Q.; Gu, Y.; Su, Y.; Chen, Y.; Chen, H.; Wang, J.; Ge, X. Summertime day-night differences of pm2.5 components (inorganic ions, oc, ec, wsoc, wson, hulis, and pahs) in changzhou, china. *Atmosphere* **2017**, *8*, 189. [CrossRef]
45. Sun, Y.; Zhang, Q.; Zheng, M.; Ding, X.; Edgerton, E.S.; Wang, X. Characterization and source apportionment of water-soluble organic matter in atmospheric fine particles (pm2.5) with high-resolution aerosol mass spectrometry and gc–ms. *Environ. Sci. Technol.* **2011**, *45*, 4854–4861. [CrossRef] [PubMed]
46. Turpin, B.J.; Huntzicker, J.J. Identification of secondary organic aerosol episodes and quantitation of primary and secondary organic aerosol concentrations during scaqs. *Atmos. Environ.* **1995**, *29*, 3527–3544. [CrossRef]
47. Wu, C.; Yu, J.Z. Determination of primary combustion source organic carbon-to-elemental carbon (oc/ec) ratio using ambient oc and ec measurements: Secondary oc-ec correlation minimization method. *Atmos. Chem. Phys.* **2016**, *16*, 5453–5465. [CrossRef]
48. Draxler, R.; Stunder, B.; Rolph, G.; Stein, A.; Taylor, A. *Hysplit4 User's Guide*; Version 4; Report; Noaa: Silver Spring, MD, USA, 2012.
49. Tao, S.; Lu, X.; Levac, N.; Bateman, A.P.; Nguyen, T.B.; Bones, D.L.; Nizkorodov, S.A.; Laskin, J.; Laskin, A.; Yang, X. Molecular characterization of organosulfates in organic aerosols from shanghai and los angeles urban areas by nanospray-desorption electrospray ionization high-resolution mass spectrometry. *Environ. Sci. Technol.* **2014**, *48*, 10993–11001. [CrossRef]
50. Riva, M.; Tomaz, S.; Cui, T.; Lin, Y.-H.; Perraudin, E.; Gold, A.; Stone, E.A.; Villenave, E.; Surratt, J.D. Evidence for an unrecognized secondary anthropogenic source of organosulfates and sulfonates: Gas-phase oxidation of polycyclic aromatic hydrocarbons in the presence of sulfate aerosol. *Environ. Sci. Technol.* **2015**, *49*, 6654–6664. [CrossRef]
51. Shakya, K.M.; Peltier, R.E. Non-sulfate sulfur in fine aerosols across the united states: Insight for organosulfate prevalence. *Atmos. Environ.* **2015**, *100*, 159–166. [CrossRef]
52. Ge, X.; He, Y.; Sun, Y.; Xu, J.; Wang, J.; Shen, Y.; Chen, M. Characteristics and formation mechanisms of fine particulate nitrate in typical urban areas in china. *Atmosphere* **2017**, *8*, 62. [CrossRef]
53. Wang, J.; Ge, X.; Chen, Y.; Shen, Y.; Zhang, Q.; Sun, Y.; Xu, J.; Ge, S.; Yu, H.; Chen, M. Highly time-resolved urban aerosol characteristics during springtime in yangtze river delta, china: Insights from soot particle aerosol mass spectrometry. *Atmos. Chem. Phys.* **2016**, *16*, 9109–9127. [CrossRef]

54. Xu, J.; Zhang, Q.; Chen, M.; Ge, X.; Ren, J.; Qin, D. Chemical composition, sources, and processes of urban aerosols during summertime in northwest china: Insights from high-resolution aerosol mass spectrometry. *Atmos. Chem. Phys.* **2014**, *14*, 12593–12611. [CrossRef]
55. Xu, J.; Zhang, Q.; Shi, J.; Ge, X.; Xie, C.; Wang, J.; Kang, S.; Zhang, R.; Wang, Y. Chemical characteristics of submicron particles at the central tibetan plateau: Insights from aerosol mass spectrometry. *Atmos. Chem. Phys.* **2018**, *18*, 427–443. [CrossRef]
56. Duan, F.; Liu, X.; He, K.; Dong, S. Measurements and characteristics of nitrogen-containing compounds in atmospheric particulate matter in beijing, china. *Bull. Environ. Contam. Toxicol.* **2009**, *82*, 332–337. [CrossRef]
57. Luo, L.; Yao, X.H.; Gao, H.W.; Hsu, S.C.; Li, J.W.; Kao, S.J. Nitrogen speciation in various types of aerosols in spring over the northwestern pacific ocean. *Atmos. Chem. Phys.* **2016**, *16*, 325–341. [CrossRef]
58. Matsumoto, K.; Yamamoto, Y.; Nishizawa, K.; Kaneyasu, N.; Irino, T.; Yoshikawa-Inoue, H. Origin of the water-soluble organic nitrogen in the maritime aerosol. *Atmos. Environ.* **2017**, *167*, 97–103. [CrossRef]
59. Matsumoto, K.; Takusagawa, F.; Suzuki, H.; Horiuchi, K. Water-soluble organic nitrogen in the aerosols and rainwater at an urban site in japan: Implications for the nitrogen composition in the atmospheric deposition. *Atmos. Environ.* **2018**, *191*, 267–272. [CrossRef]
60. Calderón, S.M.; Poor, N.D.; Campbell, S.W. Estimation of the particle and gas scavenging contributions to wet deposition of organic nitrogen. *Atmos. Environ.* **2007**, *41*, 4281–4290. [CrossRef]

Article

Spatial and Temporal Distribution of Aerosol Optical Depth and Its Relationship with Urbanization in Shandong Province

Rui Xue [1], Bo Ai [1,*], Yaoyao Lin [1], Beibei Pang [2] and Hengshuai Shang [3]

1 College of Geomrtics, Shandong University of Science and Technology, Qingdao 266590, China; xrui19@163.com (R.X.); yylinjn@163.com (Y.L.)

2 College of Earth Science and Engineering of Shandong University of Science and Technology, Qianwangang Road, Huangdao Zone, Qingdao 266590, China; berrypbb@163.com

3 Qingdao Yuehai Information Service Co., Ltd., Qingdao 266590, China; shs@oceanread.com

* Correspondence: iball@163.com; Tel.: +86-0532-8605-7285

Received: 29 January 2019; Accepted: 27 February 2019; Published: 1 March 2019

Abstract: In the process of rapid urbanization, air environment quality has become a hot issue. Aerosol optical depth (AOD) from Moderate Resolution Imaging Spectroradiometer (MODIS) can be used to monitor air pollution effectively. In this paper, the Spearman coefficient is used to analyze the correlations between AOD and urban development, construction factors, and geographical environment factors in Shandong Province. The correlation between AOD and local climatic conditions in Shandong Province is analyzed by geographic weight regression (GWR). The results show that in the time period from 2007 to 2017, the AOD first rose and then fell, reaching its highest level in 2012, which is basically consistent with the time when the national environmental protection decree was issued. In terms of quarterly and monthly changes, AOD also rose first and then fell, the highest level in summer, with the highest monthly value occurring in June. In term of the spatial distribution, the high-value area is located in the northwestern part of Shandong Province, and the low-value area is located in the eastern coastal area. In terms of social factors, the correlation between pollutant emissions and AOD is much greater the correlations between AOD and population, economy, and construction indicators. In terms of environmental factors, the relationship between digital elevation model (DEM), temperature, precipitation, and AOD is significant, but the regulation of air in coastal areas is even greater. Finally, it was found that there are no obvious differences in AOD among cities with different development levels, which indicates that urban development does not inevitably lead to air pollution. Reasonable development planning and the introduction of targeted environmental protection policies can effectively alleviate pollution-related problems in the process of urbanization.

Keywords: aerosol optical depth MODIS; urbanization; DEM; spearman rank correlation

1. Introduction

With the acceleration of China's industrialization process and the continuous growth of energy consumption, atmospheric aerosols have gradually become one of the main atmospheric pollutants [1–4], and research on China's atmospheric pollution continues to intensify. In recent years, air pollution has greatly affected people's health and daily lives. Aerosols, which play an important role in the atmospheric environment, have been a major concern of the social science community.

Atmospheric aerosol refers to a stable mixture of solid particles and liquid particles that are suspended in air. They generally have diameters between 0.001 and 100 μm. Atmospheric aerosols are mainly derived from human activities [5–9] (such as petrochemical fuel use, industrial exhaust emissions) and natural production [7,10]. Aerosols affect atmospheric visibility and human health [11–13]. Studies have shown that in cities, atmospheric aerosol concentration is closely related to chronic diseases, such as emphysema and tracheitis, and the changing trend in aerosol concentration is basically consistent with the frequency of respiratory disease development during the same period, and the two are positively correlated [13]. Atmospheric aerosols also affect the balance of atmosphere–ground radiation through direct [14–16] and indirect [10,17] climate effects, which is one of the most uncertain factors in climate change. The effects of aerosols on climate have attracted great attention from the scientific community [18,19]. The Aerosol Optical Depth (AOD), as one of the most basic optical properties of aerosols, is defined as the integral of the extinction coefficient of the medium in the vertical direction, which describes the reduction effect of aerosols on light. The MODIS AOD is a kind of data with wide coverage and more accuracy compared with other sources of currently available aerosol data, and it is also a key factor used to evaluate the degree of atmospheric pollution and to determine the aerosol climate effect [20–25]. It is often used in the study of aerosol change characteristics and regional climatic effects. In recent years, domestic and foreign scholars conducted a lot of research on the spatial and temporal distribution and variation characteristics of AOD [14,26–31], atmospheric environmental pollution monitoring [32], climate change impact [2,27], and other aspects of this topic and have made great progress.

In the process of urbanization, economic development also increases energy consumption and pollutant emissions. High-intensity human activities also cause the release of a large amount of harmful particulate matter into the atmosphere, seriously affecting air quality and human health. Studying the relationship between AOD and urbanization is of great significance for maintaining and improving the quality of the atmospheric environment [6,8–10,17,33–36].

Most of the above studies on the relationship between AOD and urbanization used population or economic factors to represent the level of urban development, but the influences of the topographic environment and climate cannot be ignored in the aerosol research process [18,37,38]. This paper uses Shandong province, China as an example to analyze the spatial and temporal distribution and change rules of AOD and its relationship with the urbanization development process, as follows: (1) The spatial and temporal distribution and change characteristics of AOD are analyzed on year, quarter and month time scales, which reveals the spatial distribution and change rules of AOD in Shandong province; (2) the correlations between AOD change and social and economic factors, such as the urban population, urban economic development data, and the main pollution emission data are analyzed to determine the degree of influence of various indicators on air quality; and (3) the correlation between AOD and urban level distribution is analyzed to study the response of AOD to urbanization. Since AOD is greatly affected by the topographic environment and climatic temperature, this paper takes into account elevation, temperature, and precipitation when studying the relationship between AOD and urbanization to better analyze and explain the correlation between them. This study can provide decision-making support for the environmental protection of cities in Shandong province [39,40], as well as providing as reference for similar analysis of cities in other provinces.

2. Study Area, Materials, and Methodology

2.1. Study Area

This study used Shandong Province as the analysis object and carried out research on the spatiotemporal distribution and variation characteristics of aerosol optical depth and its response to urbanization. Shandong Province is located on the eastern coast of China. The territory includes a peninsula and inland area. The peninsula protrudes into the Bohai Sea and the Yellow Sea. It is opposite the Liaodong Peninsula and includes Qingdao, Yantai, and Weihai, as well as most or part of Weifang, Rizhao, Dongying, and Binzhou. The inland part is bordered by Hebei, Henan, Anhui, and Jiangsu provinces from north to south, including Liaocheng, Dezhou, Jinan, Tai'an, Laiwu, Dongying, Zibo, Zaozhuang, Linyi, Jining, and the remaining parts of Weifang, Rizhao, Dongying, and Binzhou. The location map of Shandong Province is shown in Figure 1.

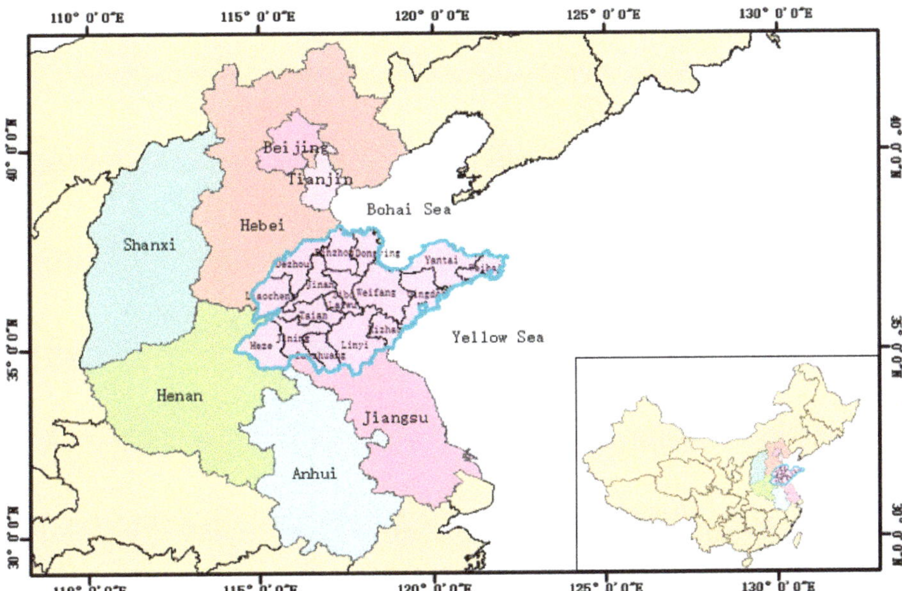

Figure 1. Shandong Province location map.

The air quality in coastal and inland cities of Shandong Province is affected by the ocean to varying degrees. The air quality in coastal cities is obviously directly controlled by the ocean [41–43]. However, the air quality in inland cities is less affected by the ocean. Generally, the wind speed in coastal areas is higher than that in inland areas, and the atmospheric diffusion conditions are good. Moreover, there are obvious differences in temperature and rainfall among cities in the province. The annual average temperature in the province decreases from southwest to northeast, and in the hilly areas of the peninsula, it is lower than in other regions. The annual average precipitation in most parts of Shandong Province is between 600 mm and 750 mm, and this has greater distribution in the north than in the south. The population density, economic strength, and urbanization levels of the cities in Shandong Province are quite different, and the urbanization development in Shandong Province is relatively unbalanced.

Based on the above analysis, when studying the spatial and temporal distribution and variation characteristics of aerosol optical data in Shandong province and their relationship with the level of urbanization, social factors such as urban population, urban construction area, and sewage discharge should be taken into account as well as the impact of the urban geographical location and climate.

2.2. Materials

The data used in this paper mainly included MODIS aerosol data from 2007 to 2017, total urban population data of Shandong Province, the green land area for urban construction, the secondary industry gross output value, urban soot, sulfur dioxide, nitrogen oxide emissions, the urban annual average temperature, precipitation, and digital elevation data.

2.2.1. Aerosol Data

AOD is defined as the integral of the extinction coefficient of the medium in the vertical direction, and it describes the effect of aerosol on light attenuation. It is one of the most important parameters of aerosols, the key physical quantity that characterizes the degree of atmospheric turbidity, and an important factor for determining the climatic effects of aerosols. The aerosol data used in this paper were derived from the MODIS c6 aerosol product released by the National Aeronautics and Space Administration (NASA) website. This product combines two inversion algorithms: the Dark Target (DT) algorithm and the Deep Blue (DB) algorithm. Many domestic scholars have verified the AOD dataset MODIS product at a wavelength of 550 nm, and AOD has been studied in some parts of China using the product. It is believed that MODIS aerosol products have reached the corresponding quality requirements and can be used to study the distribution and variation of terrestrial AOD in China [44–46]. This study used the AOD dataset with a resolution of 10 km for MODIS4, and the time series used was from 1 January 2007 to 31 December 2017.

2.2.2. Urbanization Related Data

Urbanization is a process in which the population of a region becomes relatively concentrated in cities from towns, which leads to the growth of the urban population, the expansion of urban land, and changes in the social and economic structure. This study analyzed the correlations between AOD values and the population density, the green land area for urban construction, and the regional secondary industry gross product in Shandong province, and explores the influences of these three factors on AOD value changes. During the process of urbanization, the emissions from gas pollution change continuously, which has a significant impact on AOD. This study also analyzed the relationships between AOD and soot, sulfur dioxide, and nitrogen oxides. The six elements of data were all sourced from the Shandong Provincial Bureau of Statistics. The soot, sulfur dioxide, and nitrogen oxide data are the total emissions of the index in the whole province in each year, the sources of which include industrial production as well as from factors related to people's daily lives.

Urbanization not only includes the population and social economy, but also involves cultural education, politics, transportation, and other aspects of information. It is impossible to fully understand the relationship between urbanization and AOD by relying on a single indicator. Indicators of China at the city level include the number of first-line brands, GDP, the per capita income, 211 universities, multinational top 500 entry numbers, major companies' strategic city rankings, airport throughput, large company entry, consulate number, and international routes. A total of ten indicators were calculated to comprehensively represent the development level of urbanization. In this paper, the city level data of 17 cities in Shandong Province in 2016 were used, as shown in Figure 2.

Figure 2. City level map.

2.2.3. Shandong Digital Elevation Model Data

Topography has an impact on the depth of AOD. Most scholars at home and abroad also believe that elevation has an impact on the spatial distribution of AOD. Therefore, Digital Elevation Model

(DEM) data was used as one of the influencing factors to study. DEM data of 90 m resolution were downloaded from the China natural resources satellite image cloud service platform, and the data were processed into the same resolution as AOD aerosol data by means of resampling. Figure 3 shows (a) the original DEM data and (b) the resampled DEM data.

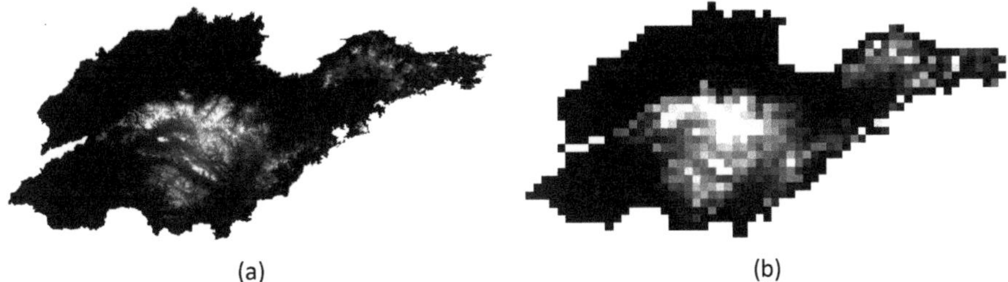

Figure 3. Digital elevation data.

2.3. Methodology

The data processing involved in this paper mainly involved the acquisition of aerosol data, projection definition, image cropping, data aggregation, and mean calculation to study the spatial and temporal distribution and variation characteristics of AOD. In this paper, first, AOD data were subjected to Lambert projection, and then the image was cut using Shandong Province as a mask. Finally, the cut AOD data were aggregated and averaged. The obtained aerosol data were first screened to ensure its accuracy before the calculation of AOD, and the valid data (data were considered valid if the percentage of valid pixels in the image data was greater than 50%) were retained. Then, the annual, quarterly, monthly aggregation, and mean values of the aerosol product data were calculated by ArcGIS. The equations are shown in (1) and (2). The monthly average AOD of Shandong Province analyzed in the paper was obtained by accumulating the daily AOD and then averaging. The quarterly average AOD and annual average AOD were obtained by accumulating the monthly average AOD and quarterly average AOD. The monthly division criterion used was the natural month when analyzing the monthly changes in AOD. The quarterly divisions used were spring from March to May, summer from June to August, autumn from September to November, and winter from December to February.

$$AOD_{QS} = \left[\sum_{s=1}^{S} (AOD(s)) \right] / S \quad (1)$$

In the equation, AOD_{QS} is the aerosol aggregated data, which was used to represent the average distribution of the aerosol optical depth, AOD represents the effective aerosol data participating in the superposition, s represents the s-th image data, and S represents the total number of superimposed images.

$$AOD_{Avg} = \left[\sum_{i=1}^{m} \sum_{j=1}^{n} (AOD(i,j)) \right] / N \quad (2)$$

AOD_{Avg} represents the average value of AOD; $AOD(i,j)$ is the pixel value of the i-th row and the j-th column in the aerosol optical depth average distribution image calculated by Equation (1); m and n respectively represent the total number of rows and the total number of columns in the image data; and N represents the total number of cells participating in the operation.

This study used the Spearman rank correlation test to measure the correlations between two variables to study the relationship between AOD values and the green land area for urban construction and the urban population. The principle was to sort the original data x_i, y_i from large to small and to record the position of the original data x_i, y_i as $\acute{x}_i, \acute{y}_i$ in the sorted list. $\acute{x}_i, \acute{y}_i$ is called the rank of x_i, y_i, and the rank difference is $d_i = \acute{x}_i - \acute{y}_i$. The Spearman rank correlation coefficient is

$$\rho_s = 1 - \frac{\sum_{i=1}^{n} d_i^2}{n^2 - 1} \quad (3)$$

The Spearman correlation coefficient was used to analyze the correlation between AOD and the urban population, which was based on a continuous time series and did not consider geographical location characteristics. However, the relationship between precipitation and AOD will change with the geographical location. This paper considered the spatial distribution of the two and finally adopted the Geographically Weighted Regression (GWR) method to explore the relationship between this factor and the urban AOD. The GWR model is shown in (4):

$$y_i = \beta_0(u_i, v_i) + \sum_{j=1}^{k} \beta_i(u_i, v_i)x_{ij} + \varepsilon_i i = 1, 2, \ldots n \quad (4)$$

In the Equation (4), x is the independent variable, y is the dependent variable, k is the number of dependent variables, j is the sample point, u_i, v_i is the spatial location of the sample point i, $\beta_o(u_i, v_i)$ is the intercept, and $\beta_i(u_i, v_i)$ is the equation coefficient, which varies with the spatial location of the sample point. Each local $\beta_o(u_i, v_i)$ is used to estimate its adjacent spatial observation value. The core of the model is the selection of the spatial weight matrix. In this study, Gaussian function was selected as the method to select the spatial weight coefficient, which is one of the most commonly used methods. The expression equation of Gaussian function is as follows:

$$W_{ij} = \frac{-1}{e^{2(d_{ij}/b)^2}} \quad (5)$$

GWR is a local form of linear regression used to model spatial variation relations. It constructs an independent equation for each element in the data set, which is used to combine the dependent variables and explanatory variables of elements falling within the bandwidth of each target element. The shape and range of the bandwidth depends on the user's input parameters, such as the kernel type, bandwidth method, distance, and number of adjacent elements. The commonly used kernel types include FIXED (the kernel surface is created by selecting the bandwidth at a certain distance) and ADAPTIVE (the kernel surface function is created according to the density of element sample distribution). The most commonly used bandwidth methods are CV, AIC, and BANDWIDTH_PARAMETER. In this study, the FIXED type was selected as the core type. The FIXED type was used to select the bandwidth according to a certain distance to create the core surface. It can generate a smoother core surface compared with ADAPTIVE. This study adopted the AIC method for bandwidth selection to achieve a better fitting degree. AIC determines the best bandwidth by the minimum information criterion, which is a standard to measure the excellent performance of statistical model fitting. The AIC function expression is as follows:

$$AIC = 2n\, ln\hat{\sigma} + n\, ln(2\pi) + n\left[\frac{n + tr(S)}{n - 2 - tr(S)}\right]. \quad (6)$$

where the trace *tr(S)* of the hat matrix S is a function of the bandwidth b, and $\hat{\sigma}$ is the maximum likelihood estimate of the variance of the random error term, and $\hat{\sigma} = \frac{RSS}{n - tr(S)}$. For the same sample data, the bandwidth corresponding to the geographic weighted regression weight function with the minimum AIC value is the optimal bandwidth. This study applied it to analyze the differences in AOD among cities with different development grades, because it is a suitable way to test the differences between the unpaired data of independent samples. The main principle of the Wilcoxon rank-sum method is to assume that there are no significant differences between the H_0 samples. The following test parameters were calculated:

$$T = \sum_{i=1}^{n} r(|d_i|)I(d_i > 0) \quad (7)$$

$$Z = \frac{T - n_1(n_1 + n_2 + 1)/2}{\sqrt{n_1 \times n_2(n_1 + n_2 + 1)/12}} \tag{8}$$

where T is the sum of the smallest rank order of the two groups of independent data. If T is in the test value interval determined by table lookup comparison, then $p > \alpha$ and H_0 is not rejected; otherwise, $p \leq \alpha$ and H_0 is rejected. The Z-statistic was used when the data volume was more than 10, and the Z-value was tested with the normal distribution method.

This study analyzed the spatial and temporal distribution of AOD in Shandong Province and its relationship with urbanization from four aspects. Firstly, the spatial and temporal distribution and the variation characteristics of AOD were analyzed by means of aggregation and mean value calculation of AOD. Secondly, the Spearman coefficient was used to analyze the correlation between AOD and urban population, the green land area for urban construction, the secondary industry GDP, and other factors to determine the impacts of urban population growth and social and economic development on AOD in the process of urbanization. Thirdly, GWR was used to explore the correlation degree of precipitation and other weather conditions on the local distribution of AOD. Fourthly, the Wilcoxon rank-sum method was used to compare and analyze the AOD differences among cities with different development levels.

3. Results and Analysis

3.1. Temporal and Spatial Distribution and Variation Characteristics of AOD

3.1.1. Annual Average Distribution and Variation Characteristics of AOD

The 11-year distribution and multi-year average distribution of AOD in Shandong Province from 2007 to 2017 are shown in Figure 4. According to the spatial distribution of AOD in the Shandong region for many years, it is concluded that from 2007 to 2017, although the intensity of the high value center of AOD in Shandong Province had some differences, the characteristics of the regional distribution center were generally of the single-high-low type and remained unchanged. A low center is located in Weihai, Yantai, and eastern Qingdao, an area close to the Yellow Sea and Bohai Sea. The AOD is around 0–0.3, which is related to the high wind speeds and good atmospheric diffusion conditions in the coastal areas. The high center is located in Liaocheng and Jining, Dezhou, Binzhou, Jinan, Tai'an, Laiwu, northern Zibo, Dongying West, and northern Weifang, which are densely populated and industrially developed, with AOD values of around 0.4–0.7.

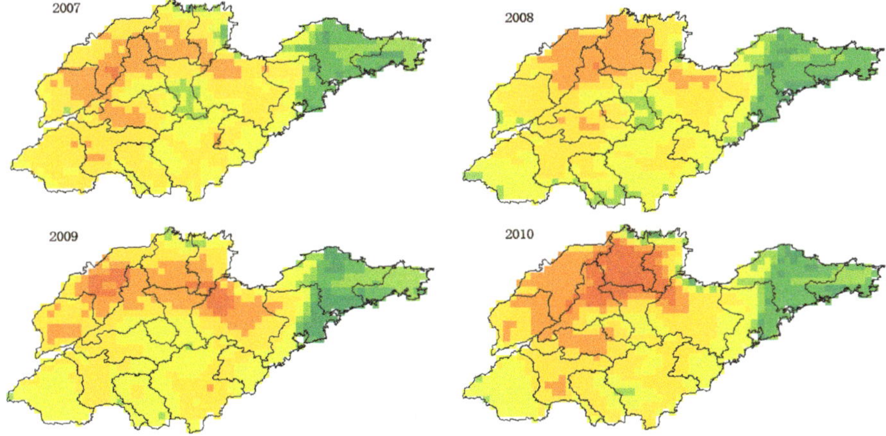

Figure 4. *Cont.*

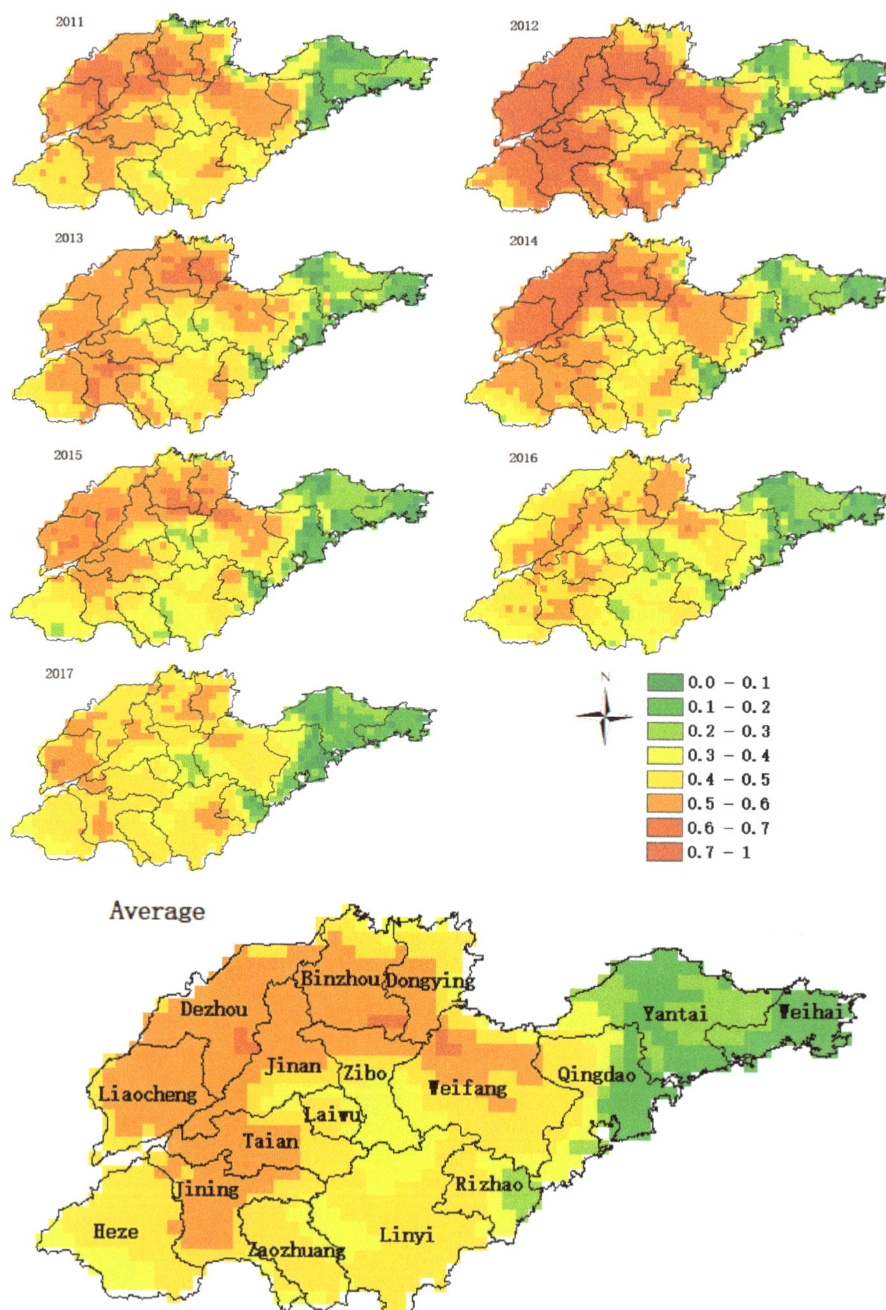

Figure 4. Aerosol Optical Depth (AOD) distribution map of Shandong Province.

The average value of the aggregated AOD in Shandong Province from 2007 to 2017 was calculated, and the AOD value changes in Shandong Province from 2007 to 2017 were obtained, as shown in Figure 5. It can be seen from the figure that, in 2007, the mean AOD of the whole province was about

0.39, slightly decreased to 0.37 in 2008, and then increased year by year until reaching its highest peak of 0.53 in 2012. It decreased significantly to 0.46 in 2013 and showed a downward trend year by year from 2014 to 2017. In general, there has been a trend of increasing first and then decreasing in the past 11 years. To further verify this trend, we constructed unitary linear regression functions for the years of 2007–2012 and 2012–2017, respectively, with 2012 as the boundary. As shown in Figure 6, 2007–2012 showed a growth trend with a slope of 0.0267 and an R^2 value of 0.7313. The years 2012–2017 showed a downward trend with a slope of −0.0263 and an R^2 value of 0.8112. From 2007 to 2012, Shandong Province made great efforts to develop its economy through industry, which brought a serious impact on the environment. Additionally, AOD showed an obvious growth trend. After 2012, Shandong Province changed its economic development model, advocated green development, dealt with polluting enterprises, and maintained a steady downward trend in AOD. The environment has improved markedly.

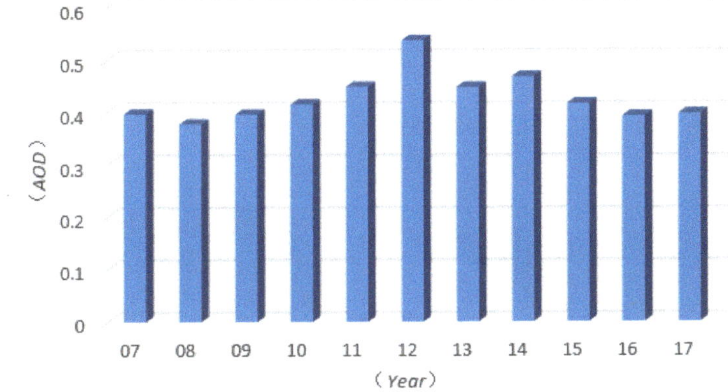

Figure 5. Annual data chart.

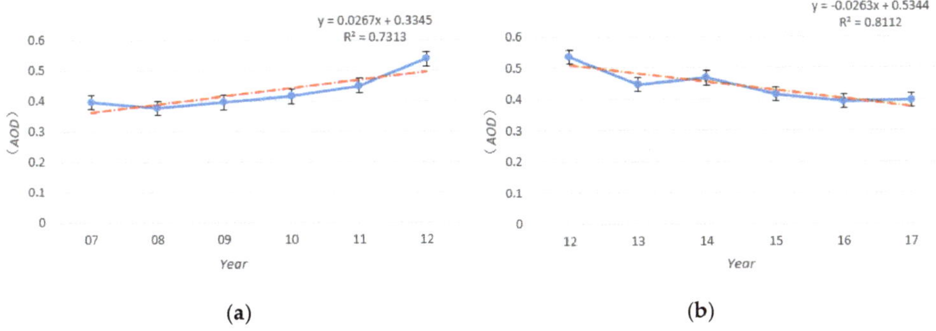

Figure 6. Linear trend chart. (**a**) Trends from 2007 to 2012, (**b**) Trends from 2012 to 2017.

3.1.2. AOD Quarterly Average Distribution and Variation Characteristics

The 11-year average AOD distribution in spring, summer, autumn, and winter in Shandong Province from 2007 to 2017 is shown in Figure 7. As can be seen from the figure, the AOD in the east of Weihai, Yantai, and Qingdao was very low, ranging from 0 to 0.3, and the change range between each quarter of the year was not large. Additionally, Dezhou, Binzhou, and Jinan were the areas with high AOD values, and AOD changed significantly among the seasons from 0.3 to 1.0. In spring and summer, the AOD in Weihai, Yantai, and eastern Qingdao was relatively low, around 0–0.3. Dezhou, Binzhou, Dongying, Jinan, the northern part of Tai'an, Laiwu, and Weifang were high AOD value areas where the AOD was about 0.6–1.0. The AOD values of Dezhou, Jinan, Binzhou, and Dongying were very high, around 0.8–1.0. In autumn, the AOD value of inland cities in Shandong Province was generally around 0.3–0.6, and in winter, the AOD value of Shandong Province was generally around 0–0.3. If the quarterly average AOD value is greater than 0.5, it is defined as a high value. The high value area of AOD was the largest in summer, followed by spring and autumn, and it was the smallest in winter. The main reasons for the significant differences in AOD between coastal and inland cities can be summarized as follows. Firstly, coastal areas mainly rely on marine resources to develop their economies, while inland areas depend on industrial development economies and therefore have serious air pollution. The hills in central Shandong have weakened the impact of monsoons from the oceans on inland areas, thereby reducing the ability of air pollutants to dissipate.

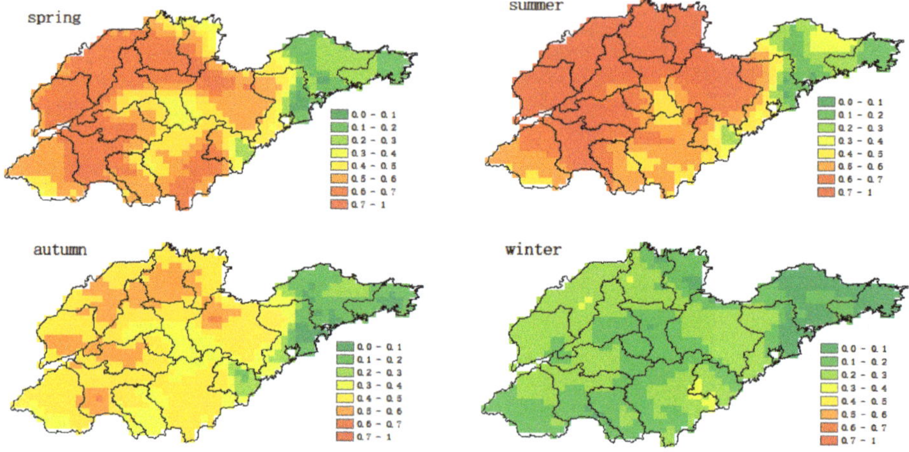

Figure 7. AOD quarterly distribution map.

The mean values of aggregated AOD in Shandong Province during the four quarters (spring, summer, autumn and winter) in 2007 to 2017 were calculated. As shown in Figure 8a, the overall trend of AOD is slowly declining, and the quarterly line changes periodically with the four seasons. The peak values varied significantly in different years, while the lowest values fluctuated less. As can be seen from the Figure 8b, the AOD value in spring in Shandong Province was around 0.35–0.7. The AOD value in summer in Shandong Province was around 0.4–0.8. The AOD value in autumn in Shandong Province was about 0.3–0.45. The AOD value in winter in Shandong province was about 0.2, the lowest value of the four seasons. There are different annual changes in different seasons. The value in autumn and winter is low and stable, and there is no obvious change trend. This is affected by the northern monsoon. The aerosol spreads well and is not easy to aggregate. The spring and summer have higher values and the trend is roughly the same as the average trend of the year. Its value is mainly affected by dust and pollution emissions. But there was an unusually low point in the summer of 2016. The reason for the abnormal AOD in the summer of 2016 is that Shandong province took compulsory measures to close down and rectify a large number of polluting enterprises in the summer of 2016, which resulted in a sharp decline in pollution in this quarter.

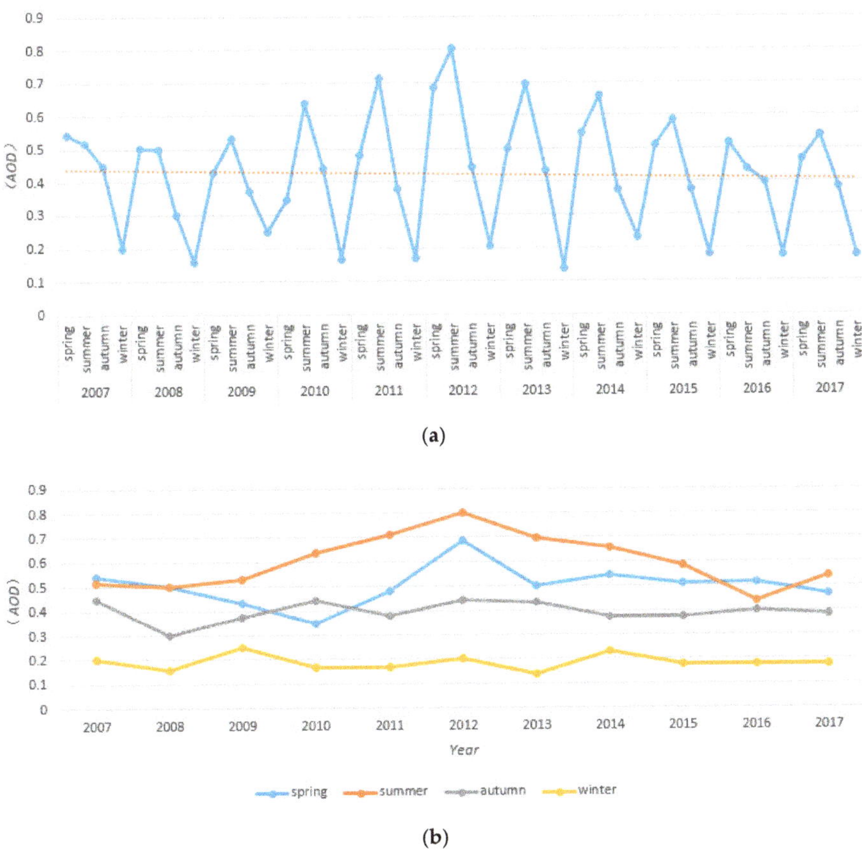

Figure 8. Quarterly data chart. (**a**) AOD overall trend chart, (**b**) Quarterly trend chart.

3.1.3. AOD Monthly Average Distribution and Variation Characteristics

The distribution of the multi-year average aerosol optical depth in Shandong Province from 2007 to 2017 is shown in Figure 9. It can be seen from the figure that the AOD in Shandong Province was generally around 0–0.3 in January, February, and December. In March, September, October, and November, except for the low AOD values of 0–0.3 in Weihai, Yantai, and eastern Qingdao, the AOD values were generally around 0.4–0.6. In April and May, the AOD values of Liaocheng, Dezhou, Binzhou, Dongying, Jinan, and Weifang in Shandong Province were about 0.4–0.8. In June, July, and August, a large number of high-value areas appeared, namely Liaocheng, Dezhou, Binzhou, Dongying, Jinan, Tai 'an, Laiwu, and Weifang, where the AOD was around 0.8–1.1. In June and July, the AOD in Binzhou and Dongying exceeded 1.1. The high value area of AOD was the largest in June and July, followed by May, then April and August, and the AOD in other months was generally less than 0.5.

The mean values of the monthly aggregated AOD in Shandong province from 2007 to 2017 were calculated, and the AOD changes in Shandong Province during the 12 months from 2007 to 2017 were obtained, as shown in Figure 10. As can be seen from the figure, the monthly variation in AOD in Shandong province presented a single peak, which increased from January onwards, and the maximum value of AOD appeared in June before gradually decreasing. This is consistent with the research results of Zheng Xiaobo et al. Due to dust invasion in spring, the AOD thickness increases. The high temperature in summer promotes the transformation of secondary organic aerosol particles to a great extent. Moreover. Shandong Province is in the peak period of wheat harvest in June each year, and people have the habit of burning straw, so a large amount of artificial aerosol particles are discharged into the atmosphere, resulting in the average aerosol optical thickness reaching the highest value during this period. However, with the increase in rainfall in July and August, the AOD decreases obviously.

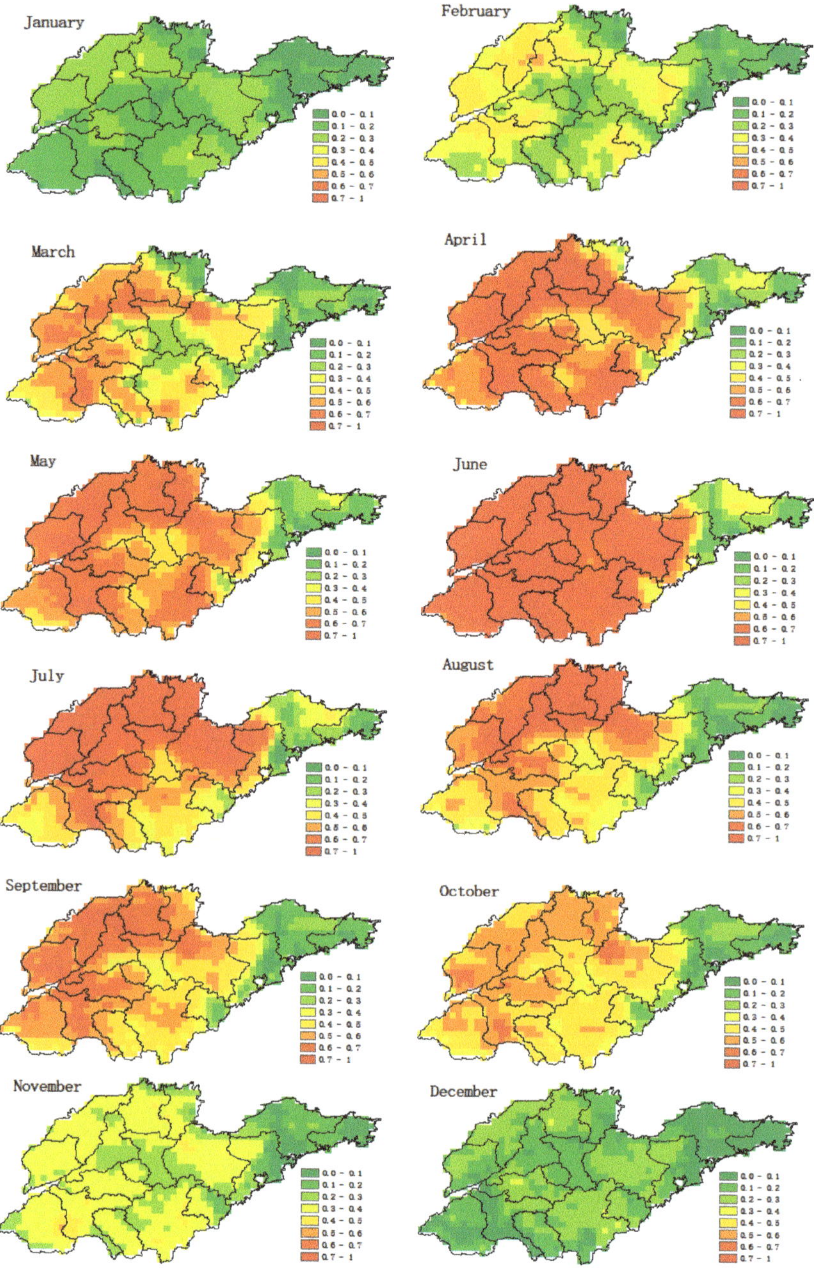

Figure 9. Monthly distribution map.

Figure 10. AOD monthly change chart.

3.2. Response Analysis of AOD to Social and Economic Factors

The AOD and socio-economic factor data from Shandong Province from 2007 to 2017 (including total urban population, green land area for urban construction, the regional GDP of the secondary industry, soot emissions, SO_2 emissions, nitrogen oxide emissions) are shown in Table 1.

Table 1. AOD and socio-economic impact factors.

Year	AOD	Population (10,000 people)	Green Land Area (ha)	Secondary Industry Regional GDP (100,000,000 Yuan)	Soot (10,000 T)	SO_2 (10,000 T)	Nitrogen Oxides (10,000 T)
2007	0.3960	3436	146,076	14,839.13	46	182	
2008	0.3769	3532	156,957	17,839.09	44	169	
2009	0.3955	3548	168,408	19,219.83	42	159	138
2010	0.4140	3839	181,055	22,163.00	39	154	141
2011	0.448	3945	188,136	24,539.45	78	183	179
2012	0.5363	4021	199,899	26,367.57	70	175	174
2013	0.4475	4130	217,366	28,163.57	70	164	165
2014	0.4673	4285	232,174	29,585.72	121	159	159
2015	0.4153	4702	240,024	30,334.56	108	153	142
2016	0.3916	4856	253,328	31,343.67	87	113	123
2017	0.3962	5024	267,944	32,942.84	54	74	116

It can be seen from Table 1 that from 2007 to 2017, the total urban population, the area of urban green land for construction, and the output value of the secondary industry in Shandong Province all showed an increasing yearly trend due to social development, and the growth rates in the past 11 years were about 7%, 83.42%, and 124.9%, respectively. From 2007 to 2017, the growth rate of green coverage in the built-up area was 83.42%. Although the green coverage rate in the built-up area showed an increasing trend, it was relatively low compared with the growth rate of the output values and the emission change rate of various pollution gases. Therefore, Spearman rank correlation test was adopted to measure the correlations between two variables, and the correlation coefficients are shown in Table 2.

Table 2. Coefficient between AOD and Socio-economic factor.

Factor	Population	Greening Land Area	Secondary Industry Regional GDP	Soot	SO_2	Nitrogen Oxides
AOD	0.191	0.21	0.21	0.393	0.329	0.867

As can be seen from Table 2, AOD has an insignificant positive correlation with the total urban population, the area of urban green land for construction, and the output value of the secondary industry, with correlation coefficients of 0.191, 0.21, and 0.21 respectively, and there were relatively high positive correlations with soot, SO_2, and nitrogen oxides, with correlation coefficients of 0.393, 0.329, and 0.867, respectively. Among these factors, there was a significant correlation with the bilateral test of nitrogen oxide emissions at the confidence level of 0.01. Generally speaking, an increase in vegetation can increase the ability of the surface to absorb particles, and then AOD will have a significant downward trend. The greening rate of urban District in Shandong Province is increasing year by year as shown in Table 1, but the inhibition effect on AOD is not obvious. Because many people are packed into cities during the process of urbanization, urban construction land expands rapidly, human activities increase significantly, producing a large amount of pollutants (soot, SO_2, nitrogen oxides, etc.) that directly acting on aerosol generation. The amount of anthropogenic aerosol is much higher than the amount of aerosol adsorbed by the vegetation.

3.3. *Response Analysis of AOD to Terrain and Weather Conditions*

The elevation data were resampled to obtain data with the same resolution as AOD, and the pixel values were extracted for the correlation analysis of the average AOD over 11 years. From 2007 to 2017, the AOD data from one year were randomly selected, and then the relevant analysis results with the average temperature and precipitation (annual precipitation depth) of prefecture-level cities in the current year were shown in Table 3.

Table 3. Coefficient between AOD and the environmental impact factor.

Factor	DEM	Temperature	Precipitation
AOD	−0.430	0.585	0.341

The correlation coefficient between elevation and AOD was −0.430, and this was a significant correlation at the 0.01 level (both sides), indicating that the altitude has a negative correlation with AOD. The higher the altitude is, the better the pollution gas diffusion will be, and the lower the AOD index will be. Places with higher altitude are not suitable for the development of heavy industry, and they is also not conducive to having a dense population, so the amount of pollution generated is also greatly reduced. The correlation coefficient between temperature and AOD was 0.585, showing a horizontal correlation at the 0.05 level (two sides). An increase in temperature promotes the generation of aerosol transformation and is not conducive to the condensation and transformation of small and medium particles in the air. The correlation coefficient between precipitation and AOD was 0.341. Generally speaking, precipitation has a scouring effect on aerosols, which is negatively correlated with AOD. However, many studies have shown that the interaction between aerosol and precipitation is complex. Precipitation is the main way to remove aerosol particles. However, the increase in aerosol optical thickness inhibits the occurrence of light rain in summer in north China, and convective precipitation is more likely to occur, and the incidence of heavy rain has increased significantly. As a result, the correlation coefficient between AOD and precipitation in the study area is positive. The Geographical Weighted Regression method was used to further analyze the influence of precipitation on AOD in local areas. In the GWR model, the precipitation data from 17 cities were used as the independent variable, and the AOD values of cities were used as the dependent variable to analyze the GWR model. As shown in Figure 11, the GWR model's fitting degree R^2 was 0.74 and the adjusted fitting degree R^2 was 0.625. The standardized residual value in 88% of the regions in the whole province ranged from −1.5 to 1.5, indicating that the GWR model had a good fitting effect overall. The urban precipitation in Shandong Province has a significant impact on the temporal and spatial distribution of AOD. In Weihai, the impact of the marine environment was shown to be greater than the adjustment of precipitation, so the residual value was obviously too small.

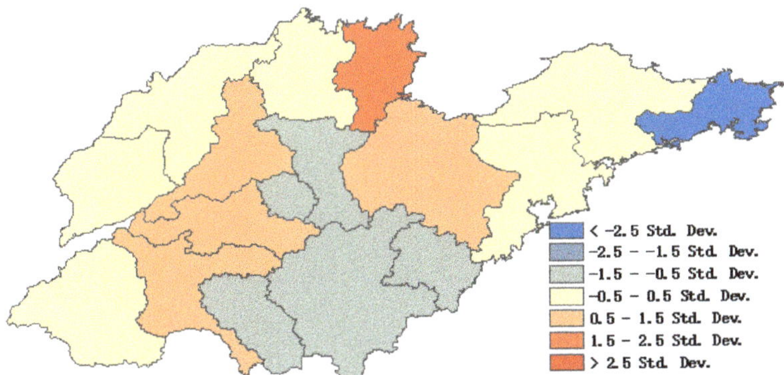

Figure 11. Geographically Weighted Regression (GWR) result graph.

3.4. Analysis of AOD's Response to the Urbanization Process

Taking the geographical distribution of AOD in Shandong Province and the spatial distribution of the city level in 2017 as an example, the relationship between AOD and urbanization level was studied. Among the 17 cities in Shandong Province in 2017, Qingdao, Jinan, Yantai and Zibo are second-tier cities with a large scale, developed economy, and frequent human activity. Weifang and Dongying are third-tier cities, with larger areas and more developed economies. Laiwu is a six-line city with a small scale, backward economic development, and a low urbanization level.

The city grade types are divided into two categories, the second-tier cities are type one, and the third-level cities and below are type two. The two groups of samples were observed using a box chart. The Figure 12 below shows that there were no significant differences in the median, range, maximum value, or minimum value between the two sets of data.

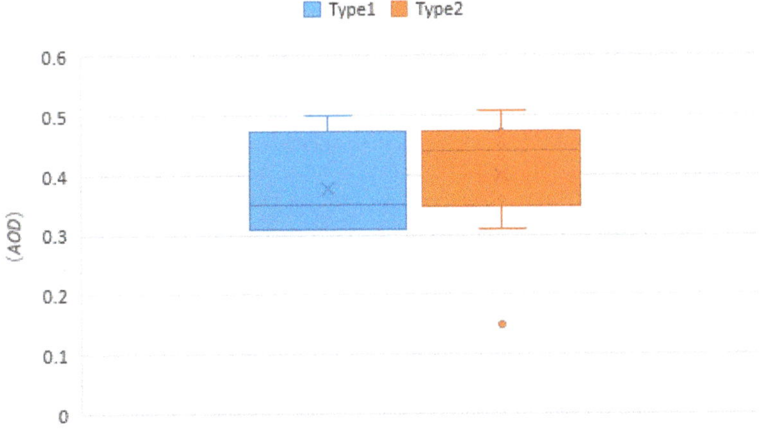

Figure 12. Box plot analysis results for different city levels.

The Wilcoxon rank-sum method was used to detect the differences between AOD data from the two city types. The sum of the order of the first class cities was 28, the sum of the order of the second class cities was 125, the Z value was –0.911, the bilateral test *p*-value was 0.362, and the exact probability test value was 0.412. These values supported the original hypothesis that there are no significant differences in AOD values between the two city types. This shows that the relationship between the urbanization development process and the current AOD values in Shandong is not obvious.

4. Conclusions

This study analyzed the spatial and temporal distribution and variation characteristics of the AOD in 17 cities on different time scales—yearly, quarterly, and monthly—by applying the daily aerosol MODIS data from Shandong Province from 2007 to 2017, as well as data on the urban population, the green land area for urban construction, the gas pollutant discharge data, the digital elevation data, the average temperature of each city, the precipitation, and the urban development level data. The Spearman coefficient was used to explore the correlations between the multi-year AOD value of Shandong Province and the urban population. GWR was used to analyze the response of AOD spatial distribution to precipitation. The Wilcoxon rank-sum method was used to verify the relationship between urbanization development level and AOD. The conclusions are as follows:

(1) From a spatial point of view, the spatial distribution of the annual average AOD in Shandong Province includes low and high centers. The low centers are Weihai, Yantai, and eastern Qingdao. The seasonal variation of AOD is relatively stable, from 0 to 0.3. One high center is concentrated in the eastern part of Liaocheng, Dezhou, Jinan, Binzhou, and Dongying, where the annual AOD is greater than 0.5. The depth of AOD in this area varies among seasons. The AOD in spring and summer increases obviously, decreases in autumn, and reaches its lowest value in winter, but this is still slightly higher than the low value of the eastern area.

(2) From a time point of view, the AOD of Shandong Province changed between 0.37 and 0.53 from 2007 to 2017, and the overall trend was an increase first, followed by a fall, and the highest value was reached in 2012. The State Council issued a notice on the comprehensive work plan for energy conservation and emission reduction in the "Twelfth Five-Year Plan" in 2011. The Shandong Provincial Government responded positively and gradually changed the economic development model and introduced many environmental protection orders to reduce the industrial pollution. AOD has shown a downward trend since 2012 with a fitting function of $y = -0.0263x + 0.5344$ and $R^2 = 0.8112$. Shandong Province has the highest AOD (0.4~0.8) in summer, followed by spring (0.35–0.7), then autumn (0.3–0.45), and finally winter (around 0.2). The monthly changes in AOD in Shandong Province are unimodal—high AOD values (greater than 0.5) appear from April to July, and the maximum values of AOD appear in June and July.

(3) There are insignificant positive correlations between AOD and the urban population, the green land area for urban construction, and the secondary industry output value. The correlation coefficients were calculated to be 0.191, 0.21, and 0.21 respectively. There are obvious positive correlations of AOD with soot, SO_2, and nitrogen oxides. The correlation coefficients were calculated to be 0.393, 0.329, and 0.867 respectively, and there was a significant correlation with nitrogen oxide emissions at the confidence level of 0.01. This shows that urban development does have a certain impact on air pollution, but pollutant emissions are the most direct cause of air quality deterioration.

(4) AOD is significantly correlated with DEM and average temperature, and the correlation coefficient with DEM was shown to be −0.430 at the significance level of 0.01 (both sides), while the correlation coefficient between temperature and AOD was calculated to be 0.585 at the significance level of 0.05 (both sides). This shows that high terrain areas are conducive to the dissipation of polluting gases. High terrain is not conducive to the development of heavy industry or dense human populations, so the pollutant gas emissions are less under normal circumstances. The higher the temperature is, the higher the AOD value is, which promotes the conversion of secondary organic aerosol particles and is not conducive to the condensation and precipitation of fine particles in the air. The spatial distribution of AOD and precipitation showed good agreement. The fitting degree of RWR of GWR model was 0.74, and the fitting degree of fit R^2 was 0.625. However, for coastal cities, the comprehensive regulation of the atmosphere by the ocean is greater than the impact of precipitation on the atmosphere, and the ocean becomes the dominant factor affecting the air quality.

(5) There is no obvious connection between AOD and the urban development level. In the process of development of different cities, economic development has different focuses; depending on resources, the economic composition is not same. Therefore, we can conclude that urban development

has a certain impact on the environment, but it can completely reduce the harm to the environment by choosing different economic development models.

This paper analyzed and summarized the spatial distribution and mean change characteristics of AOD in Shandong Province from 2007 to 2017 on annual, quarterly and monthly time scales. It is a powerful reference for understanding the temporal and spatial variations in the atmospheric aerosol optical depth in Shandong Province. It better explains the correlation between AOD and the level of urbanization development and could be a reference for urban development planning and regional atmospheric environmental governance.

Author Contributions: B.A., R.X. and Y.L. conceived and designed the research R.X. and Y.L. performed the experiments and analyzed the data; B.A., B.P. and H.S. contributed materials and analysis tools; R.X. and Y.L. wrote the paper.

Funding: This work was supported by The National Key R&D Program of China (No. 2017YFC1405004), The National Natural Science Foundation of China (No. 41401529).

Acknowledgments: We are very grateful to the editor and reviewers for their meaningful comments and helpful suggestions. This study was supported by The National Key R&D Program of China (No. 2017YFC1405004), The National Natural Science Foundation of China (No. 41401529). We are grateful to the data providers (National Aeronautics and Space Administration (NASA), Shandong Provincial Bureau of Statistics, China Natural Resources Satellite Image Cloud Service Platform) for data sources in the paper.

Conflicts of Interest: The authors declare no conflict of interest.

References

1. Cai, K.; Zhang, Q.S.; Li, S.S.; Li, Y.J.; Ge, W. Spatial-temporal variations in NO_2 and $PM_{2.5}$ over the chengdu-Chongqing Economic zone in China during 2005–2015 based on satellite remote sensing. *Sensors* **2018**, *18*, 3950. [CrossRef] [PubMed]
2. Tsai, Y.I.; Kou, S.C.; Lee, W.J.; Chen, C.L.; Chen, P.T. Long-term visibility trends in one highly urbanized, one highly industrialized, and two rural areas of Taiwan. *Sci. Total Environ.* **2007**, *382*, 324–341. [CrossRef] [PubMed]
3. Chow, J.C.; Watson, J.G.; Shah, J.J.; Kiang, C.S.; Loh, C.; Lev-On, M.; Lents, J.M.; Molina, M.J.; Molina, L.T. Megacities and atmospheric pollution. *J. Air Waste Manag. Assoc.* **2004**, *54*, 1226–1235. [CrossRef] [PubMed]
4. Murena, F. Measuring air quality over large urban areas: Development and application of an air pollution index at the urban area of Naples. *Atmos. Environ.* **2004**, *38*, 6195–6202. [CrossRef]
5. Singh, N.; Banerjee, T.; Raju, M.P.; Deboudt, K.; Sorek-Hamer, M.; Singh, R.S.; Mall, R.K. Aerosol chemistry, transport, and climatic implications during extreme biomass burning emissions over the Indo-Gangetic plain. *Atmos. Chem. Phys.* **2018**, *18*, 14197–14215. [CrossRef]
6. Semoutnikova, E.G.; Gorchakov, G.I.; Sitnov, S.A.; Kopeikin, V.M.; Karpov, A.V.; Gorchakova, I.A.; Ponomareva, T.Y.; Isakov, A.A.; Gushchin, R.A.; Datsenko, O.I. Siberian smoke haze over European territory of Russia in July 2016: Atmospheric pollution and radiative effects. *Atmos. Oceanic Opt.* **2018**, *31*, 171–180. [CrossRef]
7. Streets, D.G.; Yan, F.; Chin, M.; Diehl, T.; Mahowald, N.; Schultz, M.; Wild, M.; Wu, Y.; Yu, C.; Forschungszentrum. Anthropogenic and natural contributions to regional trends in aerosol optical depth, 1980–2006. *J. Geophysi. Res. Atmos.* **2009**, *114*, D00D18. [CrossRef]
8. Zhao, Y.; Wang, S.; Duan, L.; Lei, Y.; Cao, P.; Hao, J. Primary air pollutant emissions of coal-fired power plants in China: Current status and future prediction. *Atmos. Environ.* **2008**, *42*, 8442–8452. [CrossRef]
9. Streets, G.D.; Waldhoff, S.T. Present and future emissions of air pollutants in China: SO_2, NOx, and CO. *Atmos. Environ.* **2000**, *34*, 363–374. [CrossRef]
10. Li, Z.; Feng, N.; Fan, J.; Liu, Y.; Rosenfeld, D.; Ding, Y. Long-term impacts of aerosols on the vertical development of clouds and precipitation. *Nat. Geosci.* **2011**, *4*, 888–894. [CrossRef]
11. Sesé, L.; Nunes, H.; Cottin, V.; Sanyal, S.; Didier, M.; Carton, Z.; Israelbiet, D.; Crestani, B.; Cadranel, J.; Wallaert, B. Role of atmospheric pollution on the natural history of idiopathic pulmonary fibrosis. *Thorax* **2017**, 73. [CrossRef] [PubMed]

12. Agarwal, K.S.; Mughal, M.Z.; Upadhyay, P.; Berry, J.L.; Mawer, E.B.; Puliyel, J.M. The impact of atmospheric pollution on vitamin D status of infants and toddlers in Delhi, India. *Arch. Dis. Child.* **2002**, *87*, 111. [CrossRef] [PubMed]
13. Yang, M.; Ding, R.; Wang, S.; Shang, K. The Geological calamity and rainstorm intensity in Lanzhou city. *Arid Meteorol.* **2005**, *23*, 54–57. [CrossRef]
14. Kaiser, D.P.; Qian, Y. Decreasing trends in sunshine duration over China for 1954–1998: Indication of increased haze pollution? *Geophys. Res. Lett.* **2002**, *29*, 2042. [CrossRef]
15. Wild, M.; Gilgen, H.; Roesch, A.; Ohmura, A.; Long, C.N.; Dutton, E.G.; Forgan, B.; Kallis, A.; Russak, V.; Tsvetkov, A. From dimming to brightening: Decadal changes in solar radiation at Earth's surface. *Science* **2005**, *308*, 847–850. [CrossRef] [PubMed]
16. Zheng, X.; Kang, W.; Zhao, T.; Luo, Y.; Duan, C.; Chen, J. Long-term trends in sunshine duration over Yunnan-Guizhou Plateau in Southwest China for 1961–2005. *Geophys. Res. Lett.* **2008**, *35*, 386–390. [CrossRef]
17. Rosenfeld, D.; Dai, J.; Yu, X.; Yao, Z.; Xu, X.; Yang, X.; Du, C. Inverse relations between amounts of air pollution and orographic precipitation. *Science* **2007**, *315*, 1396–1398. [CrossRef] [PubMed]
18. Zhang, X.Y. Aerosol over China and their climate effect. *Adv. Earth Sci.* **2007**, *22*, 12–16. [CrossRef]
19. Shi, G.; Wang, B.; Zhang, H.; Zhao, J.; Tan, S.; Wen, T. The radiative and climatic effects of atmospheric aerosols. *Chin. J. Atmos. Sci.* **2008**, *32*, 826–840.
20. Ma, J.H.; Zheng, Y.F.; Zhang, H. The optical depth global distribution of black carbon aerosol and its possible reason analysis. *Scientia Meteorologica Logica Sinica* **2007**, *27*, 549–556. [CrossRef]
21. Shan, N.; Yang, X.H.; Shi, Z.J.; Yan, F. Spatial and temporal distribution of aerosol optical depth in China based on MODIS. *Sci. Soil Water Conserv.* **2012**, *10*, 24–30. [CrossRef]
22. Zeng, Q.; Chen, L.; Zhu, H.; Wang, Z.; Wang, X.; Zhang, L.; Gu, T.; Zhu, G.; Zhang, Y. Satellite-based estimation of hourly $PM_{2.5}$ Concentrations using a vertical-humidity correction method from Himawari-AOD in Hebei. *Sensors* **2018**, *18*, 3456. [CrossRef] [PubMed]
23. Aaron, V.D.; Martin, R.V.; Michael, B.; Ralph, K.; Robert, L.; Carolyn, V.; Villeneuve, P.J. Global estimates of ambient fine particulate matter concentrations from satellite-based aerosol optical depth: Development and application. *Environ. Health Perspect.* **2010**, *118*, 847–855.
24. Yang, J.M.; Qiu, J.H.; Zhao, Y.L. Validation of aerosol optical depth from terra and aqua MODIS retrievals over a Tropical Coastal Site in China. *Atmos. Oceanic Sci. Lett.* **2010**, *3*, 36–39. [CrossRef]
25. Wang, L.; Xin, J.; Wang, Y.; Li, Z.; Liu, G.; Li, J. Evaluation of the MODIS aerosol optical depth retrieval over different ecosystems in China during EAST-AIRE. *Atmos. Environ.* **2007**, *41*, 7138–7149. [CrossRef]
26. Wang, P.; Ning, S.J.; Dai, J.G.; Sun, J.M.; Lv, M.J.; Song, Q.L.; Dai, X.; Zhao, J.R.; Yu, D. Trends and variability in aerosol optical depth over North China from MODIS C6 aerosol products during 2001–2016. *Atmosphere* **2017**, *8*, 223. [CrossRef]
27. Buseck, P.R.; Pósfai, M. Airborne minerals and related aerosol particles: Effects on climate and the environment. *Proc. Natl. Acad. Sci. USA* **1999**, *96*, 3372–3379. [CrossRef] [PubMed]
28. Li, J.W.; Han, Z.W. Numerical simulation of the seasonal variation of aerosol optical depth over eastern China. *J. Remote Sens.* **2016**, *20*, 205–215. [CrossRef]
29. Zheng, X.B.; Lou, Y.X.; Chen, J. Climatology of aerosol optical depth over China from recent 10 years of MODIS remote sensing data. *Ecol. Environ. Sci.* **2012**, *32*, 265–272.
30. Zheng, X.B.; Lou, Y.X.; Zhao, T.L.; Chen, J.; Kang, W.M. Geographical and climatological characterization of aerosol distribution in China. *Scientia Geographica Sinica* **2012**, *32*, 876–883. [CrossRef]
31. Duan, Q.; Mao, J.T. Study on the distribution and variation trends of atmospheric aerosol optical depth over the Yangtze River Delta. *Acta Scientiae Circumstantiae* **2007**, *27*, 537–543. [CrossRef]
32. Su, W.H.; Shen, J.; Zhang, Q.P.; Yin, X.J.; Song, W.Z.; Lu, H.R.; Liu, J.Y.; Yuan, J.W.; Zheng, X.H.; Shen, S.Z. A study of air pollution and aerosol sulfate in the region of Tianjin. *Acta Scientiae Circumstantiae* **1982**, *2*, 329–341. [CrossRef]
33. He, L.J.; Wang, L.C.; Lin, A.W.; Zhang, M.; Xia, X.G.; Tao, M.H.; Zhou, H. What drives changes in aerosol properties over the Yangtze River Basin in past four decades? *Atmos. Environ.* **2018**, *190*, 269–283. [CrossRef]
34. Sogacheva, L.; Rodriguez, E.; Kolmonen, P.; Virtanen, T.H.; Saponaro, G.; de Leeuw, G.; Georgoulias, A.K.; Alexandri, G.; Kourtidis, K.; van der A, R.J. Spatial and seasonal variations of aerosols over China from two decades of multi-satellite observations—Part 2: AOD time series for 1995–2017 combined from ATSR ADV and MODIS C6.1 and AOD tendency estimations. *Atmos. Chem. Phys.* **2018**, *18*, 16631–16652. [CrossRef]

35. Zhao, X.; Gao, Q.; Sun, M.; Xue, Y.; Ma, R.; Xiao, X.; Ai, B. Statistical analysis of spatiotemporal heterogeneity of the distribution of air quality and dominant air pollutants and the effect factors in Qingdao Urban Zones. *Atmosphere* **2018**, *9*, 135. [CrossRef]
36. Li, L.; Chen, J.; Wang, L.; Melluki, W.; Wahid; Zhou, H. Aerosol single scattering albedo affected by chemical composition: An investigation using CRDS combined with MARGA. *Atmos. Res.* **2013**, *124*, 149–157. [CrossRef]
37. Wang, M.X. Aerosols in relation to climate change. *Clim. Environ. Res.* **2000**, *5*, 1–5. [CrossRef]
38. Sun, J.R.; Liu, Y. Possible effect of aerosols over china on east asian summer monsoon (I): Sulfate aerosols. *Adv. Clim. Chang. Res.* **2008**, *4*, 53–58. [CrossRef]
39. Jiang, Y. The Mode analysis of law on the prevention and control of atmospheric pollution: From the illegal punishment to environmental quality objective. *J. Beijing For. Uni.* **2017**, *16*, 34–42. [CrossRef]
40. Jacobson, M.Z. Atmospheric pollution—History, science, and regulation. *Phys. Today* **2003**, *56*, 65–66. [CrossRef]
41. Rowntree, P.R. The influence of tropical east Pacific Ocean temperatures on the atmosphere. *Q. J. R. Meteorol. Soc.* **2010**, *98*, 290–321. [CrossRef]
42. Deng, Z.C.; Sun, Y.L.; Zhou, H.; Song, R.X. A study on quantification of environmental capacity of marine ecology in coastal area—Qingdao city. *Marine Environ. Sci.* **2009**, *28*, 438–441. [CrossRef]
43. Jickells, T. Emissions from the Oceans to the Atmosphere, Deposition from the Atmosphere to the Oceans and the Interactions between Them. In *Global Change—The IGBP*; Springer: Berlin/Heidelberg, Germany, 2001; pp. 93–96. [CrossRef]
44. Li, C.C.; Mao, J.Y.; Liu, Q.H.; Chen, J.Z.; Yuan, Z.B.; Liu, X.Y.; Zhu, A.H.; Liu, G.Q. Using MODIS to study the distribution and seasonal variation of aerosol optical thickness in eastern China. *Chin. Sci. Bull.* **2003**, *48*, 2094–2100. [CrossRef]
45. Van, T.T.; Hai, N.H.; Bao, V.Q.; Bao, H.D.X. Remote Sensing-Based Aerosol Optical Thickness for Monitoring Particular Matter over the City. *Proceedings* **2018**, *2*, 362. [CrossRef]
46. Wei, J.; Lin, S. Comparison and evaluation of different MODIS aerosol optical depth products over the Beijing-Tianjin-Hebei region in China. *IEEE J. Sel. Topics Appl. Earth Obs. Remote Sens.* **2017**, *99*, 1–10. [CrossRef]

Article

Validation and Accuracy Assessment of MODIS C6.1 Aerosol Products over the Heavy Aerosol Loading Area

Xinpeng Tian [1,2,*] and Zhiqiang Gao [1,2]

1 CAS Key Laboratory of Coastal Environmental Processes and Ecological Remediation, Yantai Institute of Coastal Zone Research, Chinese Academy of Sciences, Yantai 264003, China

2 Shandong Key Laboratory of Coastal Environmental Processes, Yantai Institute of Coastal Zone Research, Chinese Academy of Sciences, Yantai 264003, China; zqgao@yic.ac.cn

* Correspondence: xptian@yic.ac.cn; Tel.: +86-130-2121-0988

Received: 1 August 2019; Accepted: 12 September 2019; Published: 14 September 2019

Abstract: The aim of this study is to evaluate the accuracy of MODerate resolution Imaging Spectroradiometer (MODIS) aerosol optical depth (AOD) products over heavy aerosol loading areas. For this analysis, the Terra-MODIS Collection 6.1 (C6.1) Dark Target (DT), Deep Blue (DB) and the combined DT/DB AOD products for the years 2000–2016 are used. These products are validated using AErosol RObotic NETwork (AERONET) data from twenty-three ground sites situated in high aerosol loading areas and with available measurements at least 500 days. The results show that the numbers of collections (N) of DB and DT/DB retrievals were much higher than that of DT, which was mainly caused by unavailable retrieval of DT in bright reflecting surface and heavy pollution conditions. The percentage falling within the expected error (PWE) of the DT retrievals (45.6%) is lower than that for the DB (53.4%) and DT/DB (53.1%) retrievals. The DB retrievals have 5.3% less average overestimation, and 25.7% higher match ratio than DT/DB retrievals. It is found that the current merged aerosol algorithm will miss some cases if it is determined only on the basis of normalized difference vegetation index. As the AOD increases, the value of PWE of the three products decreases significantly; the undervaluation is suppressed, and the overestimation is aggravated. The retrieval accuracy shows distinct seasonality: the PWE is largest in autumn or winter, and smallest in summer. The most severe overestimation and underestimation occurred in the summer. Moreover, the DT, DB and DT/DB products over different land cover types still exhibit obvious deviations. In urban areas, the PWE of DB product (52.6%) is higher than for the DT/DB (46.3%) and DT (25.2%) products. The DT retrievals perform poorly over the barren or sparsely vegetated area (N = 52). However, the performance of three products is similar over vegetated area. On the whole, the DB product performs better than the DT product over the heavy aerosol loading area.

Keywords: MODIS aerosol product; Collection 6.1; heavy aerosol loading; AERONET

1. Introduction

Atmospheric aerosols are small particles (0.001–100 μm) from both natural and human sources suspended in the atmosphere that can significantly influence the ecosystem [1], climate, and hydrological cycle [2] due to their effect on radiative forcing [3,4], precipitation and clouds [5]. Due to the high temporal and spatial inhomogeneity in aerosol concentrations and the complex relationship between the aerosol chemical and physical properties and cloud microphysics, the uncertainty in the estimation of the indirect aerosol forcing remains one of the highest in the climate studies today [6–8]. Satellite remote sensing is the most effective way of measuring characteristics of aerosol on the global scale. However, currently, remote sensing retrievals of aerosol properties generally achieve a low accuracy, because the radiance measured by the satellite sensor at the top of path radiance is a mixed signal [9], including not only information about the composition of the atmosphere, but also information about the Earth's surface reflectance [10].

Aerosol effects are significantly latitude dependent, i.e., at mid and lower latitudes, the aerosol loading is very strong, with consequent air pollution [11,12]. In Asia, the most substantial aerosol sources are coal, biomass burning, and dust [13,14]. The Middle East and Northern Africa are located in the region most affected by the presence of Arabian and Saharan desert dust, respectively [15]. The mass of African dust transported in the atmosphere is large [16], and it has been suggested that the transported dusts have a substantial influence on the regional radiative budget [17]. In recent years, the concentration of atmospheric particulates has increased unprecedentedly over these regions. Among them, China, India, and Pakistan's mean aerosol optical depth measured at AErosol RObotic NETwork (AERONET) sites [18] exceeds global background levels by 4–5 times [19].

Aerosol optical properties such as aerosol optical depth (AOD) were obtained from satellite sensors including the Sea-viewing Wide Field of view Sensor (SeaWiFS) [20], the Multiangle Imaging Spectroradiometer (MISR) [21], the MEdium Resolution Imaging Spectroradiometer (MERIS) [22], the Visible Infrared Imaging Radiometer Suite (VIIRS) [23], and the MODerate resolution Imaging Spectroradiometer (MODIS) [24]. MODIS sensors observe the Earth system from on board two satellites: on Terra since 1999 and on Aqua since 2002. MODIS acquires top-of-the-atmosphere (TOA) data in 36 spectral bands ranging in wavelength from 0.4 μm to 14.4 μm over both land and ocean with near-daily global coverage. The operational MODIS AOD product over land is based on two algorithms, namely, the Dark Target (DT) and Enhanced Deep Blue (DB) algorithms. Over the past decades, MODIS aerosol algorithms have experienced modifications for many times. The Collection 5 (C5) DT, C6/C6.1 DT and the DB algorithms have been extensively evaluated on both global and regional scales [25–28]. In a previous study, we provided a useful assessment of both the DT and DB aerosol products over Beijing with a specific focus on the potential of retrieval over inhomogeneous urban surfaces [28]. To date, the MODIS C6.1 aerosol products have not yet been fully and effectively verified in the polluted background of medium- and low-latitude areas with heavy aerosol loading. Thus, this study focuses more on the comparative performance of MODIS C6.1 DT, DB and the merged DT/DB AOD products over the heavy aerosol loading area.

MODIS is the first satellite observation plan designed to provide aerosol optical characteristic globally of high spatial resolution. It can provide a long time series of AOD products and can be very useful for air quality studies, etc. The objective of this study is to provide a more detailed evaluation the Terra-MODIS AOD retrieval products using the ground observations from twenty-three AERONET sites in Asia (including China, India and Pakistan), the Middle East (including United Arab Emirates, Bahrain and Kuwait) and Northern Africa (including Egypt, Niger, Benin, Mali and Nigeria) regions. For this, the Terra-MODIS C6.1 DT, DB and DB/DT AOD products (MOD04) at 10 km resolution are collected over the period 2000–2016. Then, the performance of MODIS aerosol retrieval algorithms is validated and compared against AERONET AOD measurements at the site, local and continental scales, meanwhile, the sensitivity of land use types (i.e., urban, barren or sparsely vegetation and vegetation) to aerosol and seasonal variations on aerosol retrievals are also considered and discussed.

2. Datasets and Method

2.1. AERONET Ground-Observed AOD

The AERONET (AERosol RObotic NETwork) is a worldwide network of ground stations equipped with well-calibrated Sun photometers to assess aerosol optical properties, such as their optical depth [18]. It provides a dataset of spectral AOD in the range of 0.340–1.060 μm with low uncertainty (~0.01–0.02) and high temporal resolution (every 15 min) under cloud-free conditions [29]. AERONET observations have been widely adopted for the validation of satellite-retrieved AOD [28,30,31]. For the validation of MODIS aerosol products over the heavy aerosol loading area, this study used the version 3 cloud-screened and quality-controlled level 2.0 AOD ground-based observations from 23 AERONET sites during 2000 to 2016, as shown in Figure 1, which are located in regions above the annual average WHO IT-1 value for $PM_{2.5}$ of 35 μg/m^3, and with at least 500 days of measurements. Figure 1 shows global decadal mean satellite-derived $PM_{2.5}$; this data set was inspired by work by van Donkelaar et al. [32]. $PM_{2.5}$ concentrations in large populated regions of northern India and eastern China, respectively, exceed 60 μg/m^3 and 80 μg/m^3. The summarized information of all collected sites is shown in Table 1.

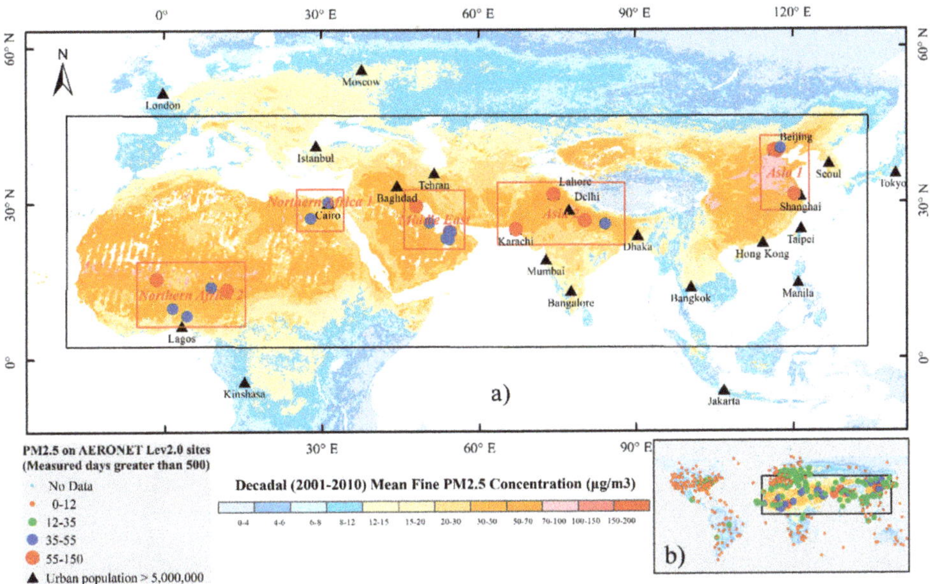

Figure 1. The geographical location of the study area. (**a**) The locations of the 23 AERONET sites are used for the evaluation of the satellite-based AOD products by circulars, and (**b**) the spatial distribution of all AERONET sites in the world. Background map is global annual average surface-level $PM_{2.5}$ concentration derived from MODIS and MISR AOD satellite data sets (2001–2010) [32].

Table 1. AErosol RObotic NETwork (AERONET) Level 2.0 data used in this study.

Region	Country	AERONET Site	Lat/Lon	Land Cover Type	Time Period	Average AOD	Average Surface Reflectance (SR)
Asia	China	Beijing (BJS)	39.977/116.381	Urban	2001–2016	0.610	0.088
		Beijing-CAMS (BJC)	39.933/116.317	Urban	2012–2016	0.611	0.079
		Taihu (THS)	31.421/120.215	Barren or Sparsely Vegetated	2005–2012	0.732	0.124
		Xianghe (XHS)	39.754/116.962	Cropland	2001–2015	0.664	0.071
		Xinglong (XLS)	40.396/117.578	Mixed Forest	2006–2012	0.271	0.037
	India	Gandhi_College (GCS)	25.871/84.128	Wetland	2006–2015	0.623	0.085
		Kanpur (KPR)	26.513/80.232	Wetland	2001–2015	0.610	0.080
	Pakistan	Karachi (KRC)	24.870/67.030	Urban	2006–2014	0.417	0.114
		Lahore (LHR)	31.542/74.325	Urban	2007–2015	0.629	0.110
Middle East	United Arab Emirates	Dhabi (DHB)	24.481/54.383	Urban	2003–2008	0.399	0.111
		Hamim (HMM)	22.967/54.300	Barren or Sparsely Vegetated	2004–2007	0.318	0.144
		Masdar_Institute (MIS)	24.442/54.617	Barren or Sparsely Vegetated	2012–2016	0.387	0.162
		Mezaira (MZR)	23.145/53.779	Barren or Sparsely Vegetated	2004–2016	0.329	0.124
		Mussafa (MSF)	24.372/54.467	Barren or Sparsely Vegetated	2004–2010	0.389	0.095
	Bahrain	Bahrain (BRN)	26.208/50.609	Urban	2000–2006	0.401	0.112
	Kuwait	Kuwait_University (KUS)	29.325/47.971	Urban	2007–2010	0.585	0.148

Table 1. *Cont.*

Region	Country	AERONET Site	Lat/Lon	Land Cover Type	Time Period	Average AOD	Average Surface Reflectance (SR)
Northern Africa	Egypt	El_Farafra (EFS)	27.058/27.990	Barren or Sparsely Vegetated	2014–2016	0.181	0.209
		Cairo_EMA_2 (CES)	30.081/31.290	Urban	2010–2014	0.355	0.089
	Niger	DMN_Maine_Soroa (DMS)	13.217/12.023	Barren or Sparsely Vegetated	2005–2010	0.493	0.112
		Zinder_Airport (ZAS)	13.777/8.990	Barren or Sparsely Vegetated	2009–2016	0.536	0.101
	Benin	Djougou (DJG)	9.760/1.599	Savannas	2004–2007	0.706	0.074
	Mali	Agoufou (AGF)	15.345/-1.479	Barren or Sparsely Vegetated	2003–2009	0.498	0.089
	Nigeria	Ilorin (ILR)	8.320/4.340	Savannas	2000–2016	0.758	0.114

Note: The SR is the surface reflectance at blue channel, the data source is MODIS surface reflectance products during 2010-2015; and the land cover type is from MODIS Land Cover product in 2013.

2.2. *Descriptions of MODIS Aerosol Algorithms and Products*

The MODIS sensor has 36 channels, ranging from 0.4 to 14.4 μm. Eight of these channels between 0.47 and 2.13 μm are used to retrieve aerosol properties over land and ocean areas by separate algorithms. The MODIS C6.1 aerosol products suite has recently been released to replace the C5 products. The Level 2 aerosol products (MOD04 for Terra and MYD04 for Aqua) are provided at a nominal spatial resolution of 10 km × 10 km at nadir. Spatiotemporally aggregated Level 3 products at 1° × 1° resolution and daily, 8-day, and monthly temporal resolution are also available. In addition to aerosol properties, the MODIS algorithm reports several diagnostic products, including a protocol for assessing the "Quality" of the retrieval known as the Quality Assurance (QA) plan [33].

2.2.1. The DT Aerosol Algorithm and Product

The C6.1 DT algorithm is based on the same algorithm principle as in C5, with subtle changes such as aerosol model, surface reflectance assumptions, cloud mask and pixel selection. First, 20 × 20 groups of pixels with a 500 m resolution at 0.47, 0.65, and 2.13 μm channels, are organized into a "retrieval box" of 10 × 10 km. All unsuitable pixels (e.g., cloud, desert, snow/ice, and inland water), and the darkest 20% and brightest 50% (at 0.65 μm surface reflectance) of pixels are discarded to reduce cloud and surface contamination. At this stage some of the brightest urban surfaces may be discarded but some pixels within the 20 × 20 kernel normally remain for computation of AOD for the 10 km pixel. Then, the surface reflectance at two visible channels can be parameterized, based on a dynamic relationship between visible channels of 0.47 and 0.65 μm and the infrared channel of 2.13 μm. The aerosol model and look-up table are employed in DT retrieval, which is conducted according to geolocation and season. Finally, the spectral AOD is obtained according to the matching values. The details of the 10 km DT algorithm were published in Levy et al. [34]. Recently, the C6.1 not only improved the process of the 10 km product, but also introduced a dataset with a 3 km spatial resolution [24]. It is also worth mentioning that a new surface reflectance relationship between shortwave infrared and visible wavelength bands were revised using a spectral surface reflectance product in the C6.1 DT algorithm [35]. The DT algorithm has been applied on the global scale, and the results show that more than 70.6% retrievals are within the estimated confidence envelope (expected error, EE) of 1 standard deviation which is $\pm(0.05 + 15\% \times AOD_{AERONET})$ [36]. In this study, the DT product at 10 km was obtained from the MODIS Level-1 and Atmosphere Archive and Distribution System (http://ladsweb.nascom.nasa.gov) for the years 2000–2016, and the "Optical Depth Land and Ocean" scientific data set (SDS) was used, with QA = 3 indicating the retrievals of highest confidence.

2.2.2. The DB Aerosol Algorithm and Product

Unlike DT, the DB algorithm was developed to retrieve aerosol properties over bright desert surfaces [37]. It performs retrievals at 1 km resolution and then aggregates pixels into a 10 km retrieval box. These pixels are masked and screened to eliminate clouds and snow/ice surfaces. For the remaining pixels, the surface reflectance is prescribed by one of several methods, dependent on location, season, and land cover type, from a global surface reflectance database in visible bands (0.412, 0.47, and

0.65 μm) at 0.1° × 0.1° resolution, which was developed by the minimum reflectivity method. Unlike the DT uncertainty estimate, the EE of the DB AOD is approximately ±(0.05 + 20% × AODAERONET), and 79% proportion of retrievals agrees within EE of the AERONET observation [26]. In this paper, the MOD04 product with the SDS of "Deep_Blue_Aerosol_Optical_Depth_550_Land_Best_Estimate" is used.

2.2.3. The Merged DT/DB Algorithm and Product

In C6 and C6.1, the new DT and DB merged AOD products (DT/DB) are based on the DT and DB AOD retrievals. The inclusion of a merged SDS was motivated by a desire to provide a more gap-filled data set than is available from the individual algorithms alone. The logic behind the merge as implemented within the C6 reprocessing was that there is a longer heritage of user familiarity with DT data over densely vegetated regions, while DB is the only data set providing coverage over arid surfaces. Three classifications were determined by monthly normalized difference vegetation (NDVI) in the merge algorithm. Over land, where NDVI ≤ 0.2 in a given month, DB data are used to populate the merged SDS, and where NDVI ≥ 0.3, DT data are used. For intermediate NDVI areas which are usually transition zones between arid and vegetated land, the algorithm whose retrieval returns the higher QA flag is used, or if both return QA = 3, the mean value is used [38]. This result is stored in the SDS named "AOD_550_Dark_Target_Deep_Blue_Combined".

2.3. Spatio-Temporal Matching for Satellite-Retrieved AOD Products with Ground-Based Observations

In this study, to match the instantaneous AOD value provided by satellites with the repeated measurements observed by AERONET, we followed the matchup methodology of Ichoku et al. [39]. The AERONET data averaged within 30 min of the MODIS overpass are extracted and compared with MODIS AOD data averaged within the 5 × 5 pixels surrounding the AERONET site. As satellite-retrieved AOD retrievals are at 550 nm, AERONET data are interpolated to 550 nm using the Angstrom exponent α, defined as:

$$\alpha = \frac{\ln(\tau_1/\tau_2)}{\ln(\lambda_1/\lambda_2)} \quad (1)$$

where τ_1, τ_2 are the AOD at wavelengths λ_1, λ_2. The nearest available pair of wavelengths from AERONET (normally 675 nm and either 440 or 500 nm) are used.

The statistical metrics considered in the evaluation are N is the number of collocated AODs; R and MR are the Pearson's correlation coefficient and the matching radio of the MODIS AOD product and the AERONET ground-observed data, respectively; RMSE, the root mean square error; MAE, the mean absolute error; RMB, the relative mean bias (the RMB > 1.0 and RMB < 1.0 indicate the overestimation and underestimation of the retrievals, respectively). PWE, PAE and PBE as percentages of collocations falling within, above and below EE envelopes. PAE and PBE represent the overestimation and underestimation of the algorithm, respectively.

$$MR = N_{(MODIS)}/N_{(AERONET)} \times 100 \quad (2)$$

$$RMSE = \sqrt{\frac{1}{n}\sum_{i=1}^{n}\left(AOD_{(MODIS)i} - AOD_{(AERONET)i}\right)^2} \quad (3)$$

$$MAE = \frac{1}{n}\sum_{i=1}^{n}\left|AOD_{(MODIS)i} - AOD_{(AERONET)i}\right| \quad (4)$$

$$RMB = \overline{AOD_{(MODIS)}}/\overline{AOD_{(AERONET)}} \quad (5)$$

$$EE = \pm(0.05 + 0.15 \times AOD_{AERONET}) \quad (6)$$

3. Results

3.1. Validation of C6.1 DT Retrievals

Table 2 provides the statistical parameters for the ground-observed and DT AOD retrievals. As shown in Table 2, we evaluate the DT retrievals at each site by analyzing the N, matching ratio (MR), R, RMSE, MAE, RMB, PWE, PAE and PBE. The MODIS C6.1 DT AODs show a high correlation (R = 0.804) with AERONET ground-observed measurements and very high accuracy (PWE = 45.6%), but have a significant overestimation over this region with high RMB (1.158, greater than 1.0) and PAE (40.7%) values. The estimation of surface reflectance (SR) is an important factor in DT algorithm over land. However, high SR values make it difficult to accomplish this discrimination as aerosol path radiance is often lower than surface radiance. It was reported by a previous study [40] that the large intercept between DT retrievals and ground-observed AOD is due to a large uncertainty in the surface reflectance estimation [41]. In the Asia region, the greatest uncertainty is observed at urban sites dominated by dust aerosols, including BJS, BJC, and LHR sites, with only 18.3%, 13.8%, and 30.6% of observations falling within the EE, respectively. This may be because the high SR (> 0.079, as shown in Table 1) of the three sites imposes a great challenge and introduces large uncertainty for the DT algorithm. There are no AOD retrievals at the KRC site with the highest SR (0.114). It is notable that the DT algorithm is not applicable in the Middle East region, where most of the land cover type is barren or sparse vegetation. The bright reflectance of dust at blue channel (SR = 0.095-0.162) limits the DT algorithm from retrieving AODs in all sites. Similar to the Middle East, the DT algorithm is only applicable to CSE, DJG and ILR sites with low reflectivity in the North Africa region. However, the N and MR are significantly smaller than those for sites in the Asian region. In addition, the DT algorithm is underestimated at DJG and ILR sites, which was attributed to overestimation of the SR and the use of inappropriate aerosol schemes in the algorithm.

Table 2. Statistical summary for validation of MOD04 DT AOD product.

Region	Site	N	MR (%)	R	PBE (%)	PAE (%)	PWE (%)	RMSE	MAE	RMB	Equation of Linear Regression
Asia	BJS	672	23.0	0.829	4.3	77.4	18.3	0.371	0.309	1.526	y = 0.876x + 0.326
	BJC	174	24.0	0.685	7.5	78.7	13.8	0.432	0.357	1.505	y = 0.637x + 0.442
	THS	52	6.3	0.863	1.9	75.0	23.1	0.390	0.322	1.424	y = 1.130x + 0.220
	XHS	917	40.8	0.927	6.9	27.3	65.8	0.247	0.145	1.106	y = 1.040x + 0.037
	XLS	240	24.9	0.871	15.0	13.3	71.7	0.128	0.080	0.969	y = 0.881x + 0.022
	GCS	558	51.7	0.821	11.1	24.4	64.5	0.192	0.132	1.071	y = 0.965x + 0.064
	KPR	1245	48.3	0.809	3.2	42.8	54.0	0.239	0.163	1.212	y = 1.040x + 0.099
	KRC	-	-	-	-	-	-	-	-	-	-
	LHR	676	57.0	0.798	4.0	65.4	30.6	0.315	0.248	1.389	y = 1.140x + 0.140
	All	4534	32.5	0.845	6.0	46.1	47.9	0.278	0.195	1.247	y = 0.990x + 0.141
Middle East	DHB	-	-	-	-	-	-	-	-	-	-
	HMM	-	-	-	-	-	-	-	-	-	-
	MIS	-	-	-	-	-	-	-	-	-	-
	MZR	-	-	-	-	-	-	-	-	-	-
	MSF	-	-	-	-	-	-	-	-	-	-
	BRN	-	-	-	-	-	-	-	-	-	-
	KUS	-	-	-	-	-	-	-	-	-	-
	All	-	-	-	-	-	-	-	-	-	-
Northern Africa	EFS	-	-	-	-	-	-	-	-	-	-
	CES	94	8.3	0.673	4.2	39.4	56.4	0.189	0.133	1.242	y = 0.906x + 0.123
	DMS	-	-	-	-	-	-	-	-	-	-
	ZAS	-	-	-	-	-	-	-	-	-	-
	DJG	243	35.0	0.861	69.2	1.6	29.2	0.290	0.234	0.682	y = 0.804x − 0.082
	AGF	-	-	-	-	-	-	-	-	-	-
	ILR	382	21.5	0.842	72.8	1.8	25.4	0.354	0.290	0.660	y = 0.782x − 0.095
	All	719	8.6	0.811	62.6	6.7	30.7	0.316	0.250	0.708	y = 0.725x − 0.012
All		5253	19.0	0.804	13.7	40.7	45.6	0.284	0.202	1.158	y = 0.915x + 0.138

Note: N is the number of collections; MR is the matching ratio of the MODIS AOD product and the ground-observed data; R is the correlation coefficient between the MODIS AOD product and the ground-observed data; PWE is the percentage within the EE; PAE is the percentage above the EE; PBE is the percentage below the EE; RMSE is the root-mean-square error; MAE is mean absolute error, it is the ratio of the satellite mean to the AERONET mean; and RMB is relative mean bias.

3.2. Validation of C6.1 DB Retrievals

Table 3 shows a comparison between the DB AOD retrievals and the AERONET ground-based observations at all sites. Compared with the DT algorithm, there are much larger numbers of matched retrievals by DB algorithm, i.e., 10258, 3755 and 4632 over sites in Asia, the Middle East, and Northern Africa, respectively. For the Asia region, the statistical data show that the MR is 73.6%, approximately 57.5% of the collections falling within the EE, with the correlation coefficient of 0.702. In this region, the overestimation of AOD is slightly greater than the underestimation (PAE = 22.6%, PBE = 19.9%). In particular, there is serious overestimation at the KRC site, for which PAE = 55.2%. The performance of the DB algorithm is significantly better than that of the DT algorithm in the Middle East region, where the MR is 70.6%, with 47.3% of the collections falling within the EE. In the Northern Africa region, the DB AOD retrievals show a high agreement with AERONET AOD measurements and approximately 49.2% of the retrievals falling within the EE, with low RMSE (0.248) and MAE (0.166) errors. On the whole, the DB algorithm is significantly improved compared to the DT algorithm in this study, i.e., the RMSE and MAE errors are decreased by 14.5% and 20.8%, respectively. Additionally, approximately 53% of the DB retrievals fall within the EE. Although DB was slightly underestimated in comparison to DT, the overestimation of DB retrievals was significantly improved, with the PAE decreased by 63%. This indicates that the DB product is recommended for aerosol applications in this study.

Table 3. Statistical summary for validation of MOD04 DB AOD product.

Region	Site	N	MR (%)	R	PBE (%)	PAE (%)	PWE (%)	RMSE	MAE	RMB	Equation of Linear Regression
Asia	BJS	2462	84.1	0.895	17.1	19.7	63.2	0.261	0.148	1.011	y = 0.929x + 0.043
	BJC	532	73.4	0.864	17.1	22.7	60.2	0.303	0.169	1.051	y = 0.972x + 0.039
	THS	370	45.1	0.799	26.2	20.3	53.5	0.321	0.203	1.002	y = 1.010x − 0.006
	XHS	1878	83.5	0.911	8.7	36.6	54.7	0.281	0.173	1.133	y = 0.986x + 0.088
	XLS	562	58.4	0.796	20.4	13.9	65.7	0.169	0.098	0.933	y = 0.798x + 0.034
	GCS	842	78.0	0.740	22.4	22.8	54.8	0.290	0.188	1.039	y = 1.060x − 0.012
	KPR	1865	72.3	0.752	14.0	23.9	62.1	0.250	0.157	1.065	y = 0.934x + 0.080
	KRC	836	59.4	0.603	55.2	5.9	38.9	0.227	0.169	0.672	y = 0.462x + 0.087
	LHR	911	76.8	0.773	27.0	20.1	52.9	0.229	0.164	0.936	y = 0.825x + 0.066
	All	10258	73.6	0.862	19.9	22.6	57.5	0.261	0.161	1.020	y = 0.958x + 0.034
Middle East	DHB	230	74.2	0.558	41.8	24.3	33.9	0.235	0.183	0.900	y = 0.625x + 0.111
	HMM	581	87.5	0.765	16.0	16.5	67.5	0.137	0.094	1.012	y = 0.859x + 0.048
	MIS	670	70.5	0.506	34.6	28.2	37.2	0.238	0.178	0.969	y = 0.542x + 0.170
	MZR	1274	86.4	0.800	11.6	34.0	54.4	0.158	0.114	1.156	y = 0.957x + 0.065
	MSF	470	59.9	0.585	38.5	28.9	32.6	0.254	0.195	0.992	y = 0.626x + 0.146
	BRN	193	30.9	0.626	44.6	17.1	38.3	0.231	0.177	0.809	y = 0.555x + 0.106
	KUS	337	65.6	0.776	16.9	42.4	40.7	0.307	0.222	1.185	y = 0.784x + 0.215
	All	3755	70.6	0.702	23.8	28.9	47.3	0.210	0.150	1.047	y = 0.769x + 0.104
Northern Africa	EFS	342	55.2	0.659	7.9	63.2	28.9	0.235	0.189	1.614	y = 0.675x + 0.214
	CES	934	82.8	0.651	29.3	26.2	44.5	0.180	0.137	0.951	y = 0.604x + 0.120
	DMS	647	58.1	0.769	27.0	22.6	50.4	0.214	0.149	0.951	y = 0.706x + 0.104
	ZAS	604	40.0	0.887	19.5	27.0	53.5	0.241	0.156	1.076	y = 1.070x + 0.004
	DJG	395	56.8	0.839	40.8	9.6	49.6	0.257	0.185	0.846	y = 0.807x + 0.027
	AGF	1089	72.4	0.812	20.5	27.2	52.3	0.285	0.172	1.065	y = 0.968x + 0.047
	ILR	621	34.9	0.855	17.4	26.2	56.4	0.304	0.203	1.030	y = 0.951x + 0.066
	All	4632	55.5	0.837	23.4	27.4	49.2	0.248	0.166	1.021	y = 0.903x + 0.058
ALL		18,645	67.5	0.847	21.6	25.0	53.4	0.248	0.160	1.025	y = 0.931x + 0.047

3.3. Validation of C6.1 DT/DB Retrievals

To evaluate the performance, a total of collocated 11,537 MODIS DT/DB aerosol retrievals were compared with AERONET ground-based measurements. As shown in Table 4, it is easy to find that AOD retrievals exhibit overall high correlations with AERONET AOD measurements (R = 0.841), with 53.1% of them falling within the EE, indicating good performances over the heavy aerosol loading areas. However, the accuracies vary greatly at different sites, where only five sites had more than 60% of the collections falling within the EE. In the Asia region, the DT/DB AOD retrievals have an underestimation of 9.8% and an overestimation of 36.5%; the PWE is 53.7%. Similar to the DB algorithm, the DT/DB algorithm also underestimates at the KRC site, where PBE = 48.1%. However, the opposite trend is observed at sites in the North Africa region, where there is an average underestimation of 35.2% and an overestimation of 17.8%, with 47.0% of the collections falling within the EE. Compared with the DT algorithm, the DT/DB algorithm has improved in this region, but this is still not applicable at the DHB, MSF and BRN sites. For all sites, the DT/DB AOD retrievals achieved high correlation (R = 0.841) with AERONET ground measurements, with low RMSE (0.238) and MAE (0.157) errors. However, the DT/DB algorithm has an obvious overestimation in the three regions (PAE = 30.3%, RMB = 1.082).

Table 4. Statistical summary for validation of MOD04 DT/DB AOD product.

Region	Site	N	MR (%)	R	PBE (%)	PAE (%)	PWE (%)	RMSE	MAE	RMB	Equation of Linear Regression
Asia	BJS	1742	59.5	0.853	9.0	38.7	52.3	0.271	0.176	1.194	y = 0.879x + 0.143
	BJC	347	47.9	0.800	7.8	50.1	42.1	0.303	0.213	1.294	y = 0.879x + 0.175
	THS	63	7.7	0.883	4.8	63.5	31.7	0.022	0.004	1.381	y = 1.250x + 0.087
	XHS	1418	63.1	0.917	5.9	33.8	60.3	0.259	0.154	1.153	y = 1.050x + 0.056
	XLS	314	32.6	0.879	16.9	11.8	71.3	0.121	0.077	0.945	y = 0.889x + 0.013
	GCS	684	63.3	0.768	14.6	23.7	61.7	0.208	0.141	1.043	y = 0.913x + 0.078
	KPR	1469	57.0	0.788	6.1	37.9	56.0	0.232	0.157	1.160	y = 0.968x + 0.113
	KRC	285	20.3	0.603	48.1	4.2	47.7	0.189	0.134	0.679	y = 0.335x + 0.124
	LHR	684	57.7	0.779	5.6	61.4	33.0	0.309	0.240	1.356	y = 1.110x + 0.138
	All	7006	50.3	0.856	9.8	36.5	53.7	0.254	0.167	1.164	y = 0.980x + 0.094
Middle East	DHB	-	-	-	-	-	-	-	-	-	-
	HMM	542	81.6	0.814	17.2	12.7	70.1	0.113	0.083	0.956	y = 0.832x + 0.039
	MIS	30	3.2	0.521	36.7	20.0	43.3	0.167	0.137	0.885	y = 0.687x + 0.072
	MZR	1149	78.0	0.821	11.9	31.2	56.9	0.144	0.105	1.126	y = 0.947x + 0.057
	MSF	-	-	-	-	-	-	-	-	-	-
	BRN	-	-	-	-	-	-	-	-	-	-
	KUS	37	7.2	0.884	24.4	27.0	48.6	0.332	0.228	1.007	y = 0.849x + 0.123
	All	1758	33.0	0.834	14.2	25.2	60.6	0.142	0.101	1.066	y = 0.906x + 0.052
Northern Africa	EFS	116	18.7	0.415	12.1	52.6	35.3	0.205	0.165	1.414	y = 0.349x + 0.227
	CES	337	39.9	0.672	32.6	22.6	44.8	0.167	0.127	0.940	y = 0.808x + 0.043
	DMS	469	42.1	0.783	29.2	19.0	51.8	0.207	0.139	0.901	y = 0.654x + 0.097
	ZAS	442	29.3	0.910	20.3	22.9	56.8	0.413	0.137	1.040	y = 1.080x − 0.019
	DJG	267	38.4	0.872	63.7	2.6	33.7	0.278	0.221	0.710	y = 0.800x − 0.061
	AGF	760	50.5	0.841	23.3	20.1	56.6	0.228	0.138	0.961	y = 0.859x + 0.044
	ILR	382	21.5	0.842	72.8	1.8	25.4	0.354	0.290	0.660	y = 0.782x − 0.095
	All	2773	33.2	0.823	35.2	17.8	47.0	0.243	0.167	0.870	y = 0.784x + 0.041
ALL		11,537	41.8	0.841	16.6	30.3	53.1	0.238	0.157	1.082	y = 0.934x + 0.073

3.4. Comparison of DT, DB and DT/DB Products with Ground-Observed Data in Each AOD Bin

Figure 2 shows the scatter plots of MODIS DT, DB and DT/DB AOD retrievals against AERONET measurements for the years 2000–2016 in the 0 to 4.0 AOD bins. The number of collections for DT retrievals is lower than that for DB and DT/DB retrievals because the DT algorithm is unable to retrieve AOD over bright and complex surface types. In the correlation comparison between satellite retrieval and ground observation, the slope of the linear regression is linked to systematic uncertainties such as assumptions of aerosol models, while the constant offset tends to be associated with factors such as deviation in surface reflectance estimation [26,42]. For DT retrievals, the slope is lower than DB and DT/DB and the intercept is higher than both. As shown in Figure 3, the PWE values of the three products are decreased significantly with increasing AOD. It is worth mentioning that the PWE value of DT algorithm is 51.1% when AOD > 2.0, which is due to the number of collections being very small (N = 47). In addition, the PAE and PBE values decreased and increased, respectively, indicating that overestimation was significantly suppressed, and underestimation increased with the increase of AOD. The DT retrievals have lowest PWE value, highest overestimation, and lowest underestimation in each AOD bin, except for AOD > 2.0. The DB and DT/DB retrievals have lower uncertainty when

AOD < 0.25 with PAE = 59.3% and 61.1%, respectively. However, the DB and DT/DB algorithms exhibit underestimation during polluted days (AOD > 1.0), approximately 30% of the collections above EE.

On the whole, the accuracy of the DT, DB and DT/DB algorithms over the heavy aerosol loading and high surface reflectance areas is lower than on the global scale. The verification results show that none of the aerosol products are suitable for application in research on atmospheric aerosols under heavy aerosol loading; thus, there is important theoretical and practical significance for proposing a high-precision method of aerosol retrieval.

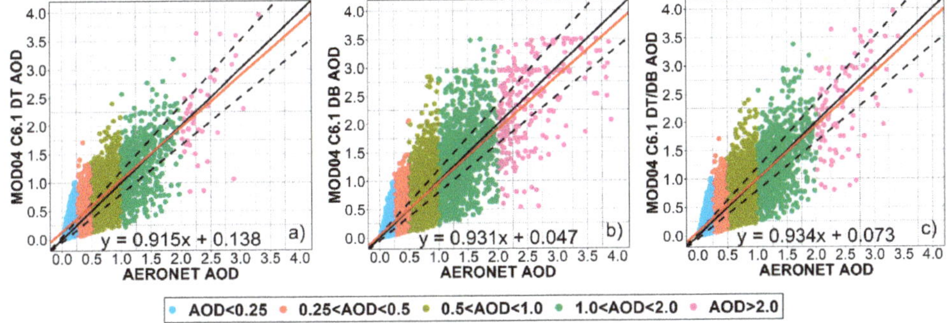

Figure 2. Validation of MOD04 C6.1 DT (**a**), DB (**b**), and combined DT/DB (**c**) AOD retrievals at 10 km resolution against AERONET measurements for the years 2000–2016. The dashed lines = EE lines, black solid line = 1:1 line, and red solid line = regression line.

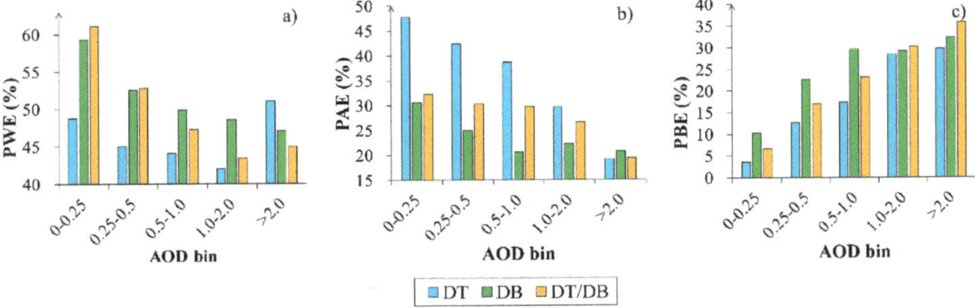

Figure 3. The percentage within in (PWE), above (PAE) and below (PBE) the expected error for the DT (**a**), DB (**b**), and DT/DB (**c**) products in each AOD bin over heavy aerosol loading area for the years 2000–2016.

3.5. Seasonal Differences of DT, DB and DT/DB Products

Previously, strong seasonality has been noted in the performance of Aqua MODIS products (including DT, DB and merged DT/DB) [43], which instigated us to examine the seasonality of Terra MODIS AOD. Figures 4 and 5 summarize the seasonal variation of retrieval accuracy for Terra MODIS and AERONET comparison of AOD at 550 nm, indicating that the statistical parameters of DT, DB and DT/DB aerosol retrievals exhibit strong seasonality and similar variation trends. The number of collections for three products is the greatest in spring, moderate in autumn and winter, and the smallest in summer. The value of PWE is highest in autumn, moderate in spring and winter, and lowest in summer. The variation of overestimation is similar, the largest in summer, the smallest in winter for DT and DT/DB products, but the smallest overestimation of DB product is in autumn. The PBE values of the three products are lower than those of PAE in different seasons, indicating that the underestimation of the aerosol retrievals is weaker than overestimation. The underestimation also has strong seasonality, the PBE values of DT and DT/DB are highest in winter and lowest in summer, while the PBE values of DB product are the opposite. In the four seasons, the number of collections of DB product is more than DT and DT/DB products. The accuracy of DB retrievals is superior to that of DT retrievals, due to the more accurate estimation of the surface reflectance. DT has the lowest accuracy, the largest overestimation and the minimum collections, which further demonstrates that the DT algorithm is not suitable for areas with high surface reflectance and pollution. We analyzed the single scattering albedo (SSA) value at Beijing AERONET site from 2001 to 2016 and found that the average values are 0.90, 0.92 at 440, 676 nm, respectively. For the study area, the high setting of the SSA value in the MODIS lookup table may be the one of the reasons for the uncertainty of AOD retrieval.

Figure 5 summarizes the error statistics for DT, DB, and DT/DB products against AERONET AOD ground-observed measurements in the four seasons, which indicates that the performance of the DB and DT/DB algorithms is slightly better than the DT algorithm in each season. Figure 5 also illustrates that three products show a strong seasonality when retrieving AOD. Correlations of the DB, DT/DB retrievals and AERONET AOD ground-observed measurements are up to approximately 0.80 in any season, but the correlation coefficient between the MODIS DT retrievals and the AERONET ground-observed AODs in winter decreases to 0.703 as shown Figure 5. Most values of the RMSE are much less than 0.30 in the four seasons for the three products, except DT in winter. The MAE error is higher in summer than other seasons. The RMB errors are greater than 1.0 in the four seasons, which indicates overestimation of the retrievals, except DT in winter, as shown in Figure 3, where PBE = 32.3%. In addition, the values of RMSE and MAE errors for DT product are larger than DB and DT/DB products and the RMB errors for DB and DT/DB products are close to 1.0. In general, the MODIS DT product shows poor ability to retrieve AOD as it has a low PWE of 37.4–50.5% with high RMSE of 0.233–0.380 and MAE of 0.166–0.283. The reason for seasonal variation in accuracy at these sites may be due to monsoon and downwind biomass burning sources [44,45].

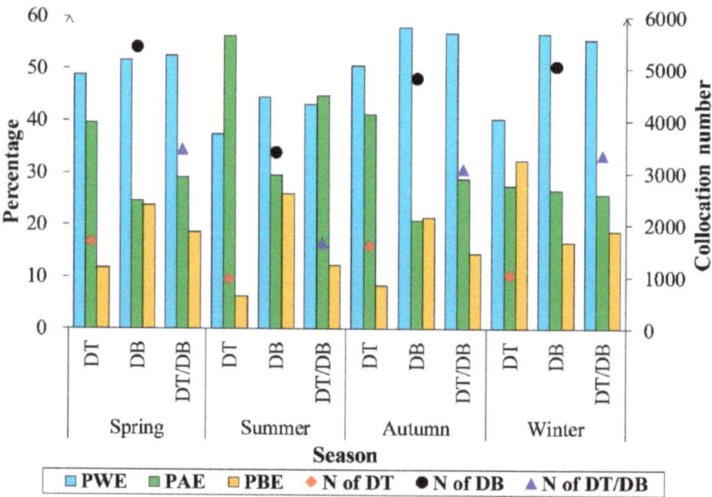

Figure 4. The percentage within in (PWE), above (PAE) and below (PBE) the expected error and the number of collocations for the DT, DB and DT/DB products in the four seasons over heavy aerosol loading area for the years 2000–2016.

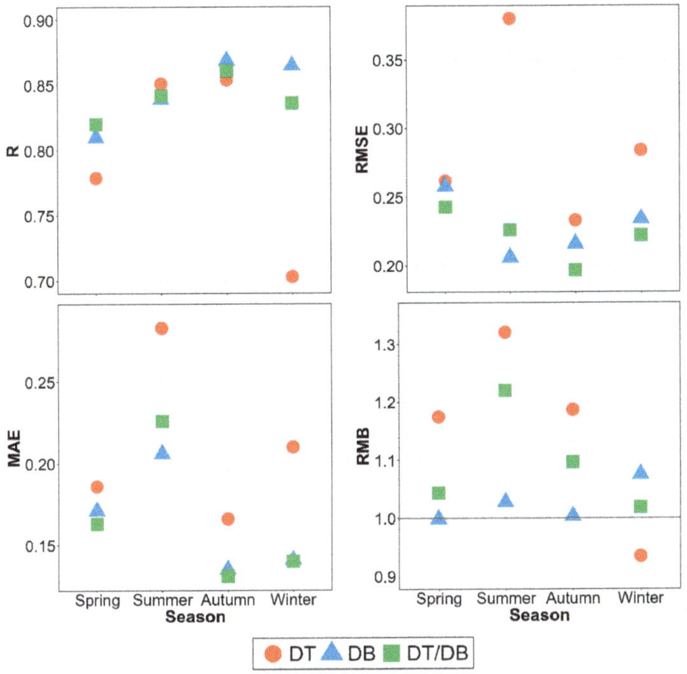

Figure 5. Error statistics for DT, DB, and DT/DB products against AERONET AOD ground-observed measurements in the four seasons.

3.6. *Adaptability of DT, DB and DT/DB Products over Different Land Cover Types*

Due to the diurnal variation of the aerosol sources, components and particle size distributions, assessing the availability of DT, DB and DT/DB products over various land cover types is becoming imperative. To explore the adaptability of three products over different land cover types, in this study, the 23 AERONET sites are divided into three land cover types: (1) urban (8 sites: BJS, BJC, KRC, LHR, DHB, BRN, KUS, and CES); (2) barren or sparsely vegetated (9 sites: THS, HMM, MIS, MZR, MSF, EFS, DMS, ZAS, and AGF); and (3) vegetated (6 sites: XHS, XLS, GCS, KPR, DJG, and ILR), according to the MCD12Q1 Land Cover product in 2013 year and the MOD09A1 8-day composited Surface Reflectance product from 2010 to 2015. Figure 6 illustrated the numbers of collections and the value of PWE, PAE and PBE for three products over urban, barren or sparsely vegetated and vegetated areas and Figure 7 shows the error statistics.

There were totals of 1616, 6435 and 3432 MODIS/AERONET collections over urban area for the DT, DB and DT/DB products, respectively. It is easy to find that the DB retrievals have the highest PWE, the lowest PAE, high correlations with AERONET AOD measurements, and small RMSE and MAE errors, with a value of RMB close to 1.0. For DT retrievals, only 25.2% of the retrievals falling within the EE had large RMSE and MAE errors. The DT product has a higher PAE than the DB product, indicating that DT retrievals seriously overestimated aerosol loading. The DT/DB retrievals have a good correlation, small RMSE and MAE errors, with 46.3% of the collections falling within the EE. However, the DT/DB results exhibited an overestimation, the PAE is slightly higher than that for the DB product. The accuracy of DT/DB retrievals over urban areas is better than the DT product and worse than the DB product. On the whole, the DB retrievals have better applicability in urban areas. Unlike DT and DT/DB products, the DB retrievals are somewhat underestimated over this area.

For barren or sparsely vegetated areas, the collections N for the DT algorithm are much less than other algorithms over this area, with only a small amount at the THS site (N = 52, as shown in Table 2). The results show that the DT product has poor ability to retrieve AOD over barren or sparsely vegetated area. The accuracies of the DB and DT/DB products are similar, with 49.7% and 56.9% of the collections falling within the EE, respectively. However, the collections for DB retrievals with AERONET are almost 1.7 times those for DT/DB product.

We have collected 3432, 3571 and 4534 collections for DT and DB and DT/DB products over the vegetated area. Figures 6 and 7 show that the vegetated sites have highest PWE value for each product, with less overestimation, high correlations, small RMSE and MAE errors, which is not surprising, as the DT algorithm is tuned to vegetated targets.

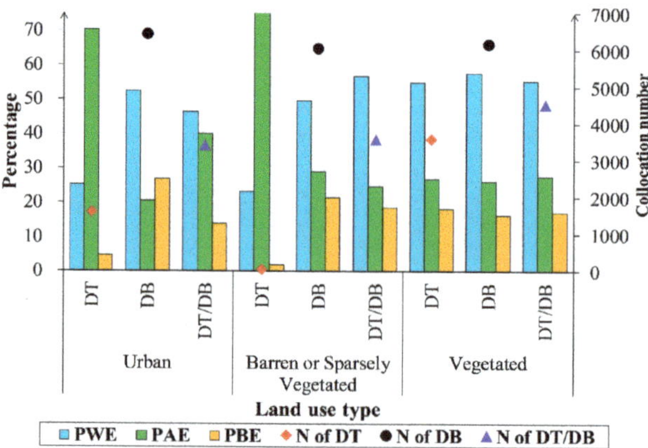

Figure 6. The percentage within in (PWE), above (PAE) and below (PBE) the expected error and the number of collocations for the DT, DB and DT/DB products over different land cover types for the years 2000–2016.

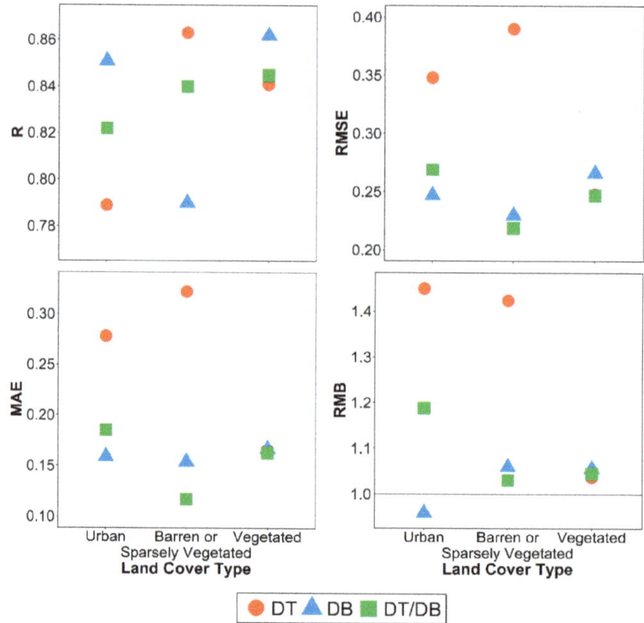

Figure 7. Error statistics for DT, DB, and DT/DB products against AERONET AOD ground-observed measurements over different land cover types.

4. Discussion and Conclusions

The objective of this study was to evaluate the Terra-MODIS C6.1 AOD retrieval product including Dark Target, Deep Blue and the combined DT/DB product at 10 km spatial resolution using ground-observed measurements from AERONET 23 sites over heavy aerosol loading areas during the period 2000 to 2016. The results showed the poorest performance for the DT algorithm over the study area, with only 45.6% of the collections falling within the EE, and with larger RMSE and MAE errors than the DB and DT/DB retrievals. The DB AOD retrievals were in good agreement with AERONET AOD measurements (R = 0.847) at most of the sites, and 53.4% of retrievals fall within in the EE with low RMSE and MAE errors, but the AOD was significantly underestimated when AOD > 1.0. The combined DT/DB AOD appeared to be superior to DT AOD due to the greater contribution of DB retrievals over bright-reflecting source regions. On the whole, over medium- and low-latitude land areas, the value of PWE for all algorithms was lower than the value on a global scale.

In general, the accuracy of the MODIS aerosol product is related not only to the land cover type but also to seasonal variation. For the urban and barren or sparsely vegetation areas, the accuracy of DB and DT/DB algorithms is similar, and better than that of DT retrievals. The performance of three products is similar over vegetation area. The accuracy of DT product is not significantly better than that of DB and DT/DB products, which may be related to both underestimation of surface reflectance and inappropriate use of an absorbing aerosol model in the study area. For the all products, the value of PWE is the largest in autumn, and the smallest in summer, overall, the performance of DB and DT/DB are better than DT in most seasons.

This study provided an overview and initial analysis of the three typical products of Terra-MODIS C6.1 over medium- and low-latitude land areas, where there is heavy aerosol loading and increasing air pollution. The analysis has shown that the DB algorithm has the best performance in this area, and while not optimal, the current merged DT/DB algorithm nevertheless does provide a data set which will be suitable for quantitative scientific applications.

Author Contributions: Conceptualization, X.T. and Z.G.; methodology, X.T. and Z.G.; data curation, X.T.; formal analysis, X.T.; validation X.T.; funding acquisition, Z.G.; project administration, Z.G.; supervision, Z.G.; writing of the original draft, X.T.; writing of review and editing, Z.G.

Funding: This work was supported by the National Natural Science Foundation of China (NSFC) fund project (41876107), NSFC-Shandong joint fund project (U1706219), Basic Special Program of Ministry of Science and Technology (2014FY210600) and Aoshan Science and Technology Innovation Program of Qingdao National Laboratory for Marine Science and Technology (2016ASKJ02).

Acknowledgments: The data for this paper are available at Level-1 and Atmosphere Archive & Distribution System (LAADS) Distributed Active Archive Center (DAAC) (http://ladsweb.nascom.nasa.gov) and AERONET Web (http://aeronet.gsfc.nasa.gov). The authors would like to thank GSFC and AERONET for kindly providing the data. In addition, a very special acknowledgement is made to the editors and referees who provided important comments that improved this paper.

Conflicts of Interest: The authors declare no conflict of interest.

References

1. Grantz, D.A.; Garner, J.H.B.; Johnson, D.W. Ecological effects of particulate matter. *Environ. Int.* **2003**, *29*, 213–239. [CrossRef]
2. Ramanathan, V.; Crutzen, P.J.; Kiehl, J.T.; Rosenfeld, D. Atmosphere-Aerosols, climate, and the hydrological cycle. *Science* **2001**, *294*, 2119–2124. [CrossRef] [PubMed]
3. Kaufman, Y.J.; Tanre, D.; Boucher, O. A satellite view of aerosols in the climate system. *Nature* **2002**, *419*, 215–223. [CrossRef] [PubMed]
4. Lolli, S.; Campbell, J.R.; Lewis, J.R.; Gu, Y.; Marquis, J.W.; Chew, B.N.; Liew, S.C.; Salinas, S.V.; Welton, E.J. Daytime Top-of-the-Atmosphere Cirrus Cloud Radiative Forcing Properties at Singapore. *J. Appl. Meteorol. Clim.* **2017**, *56*, 1249–1257. [CrossRef]

5. Rosenfeld, D.; Andreae, M.O.; Asmi, A.; Chin, M.; de Leeuw, G.; Donovan, D.P.; Kahn, R.; Kinne, S.; Kivekas, N.; Kulmala, M.; et al. Global observations of aerosol-cloud-precipitation-climate interactions. *Rev. Geophys.* **2014**, *52*, 750–808. [CrossRef]
6. Lohmann, U.; Feichter, J. Global indirect aerosol effects: a review. *Atmos. Chem. Phys.* **2005**, *5*, 715–737. [CrossRef]
7. Chen, Y.; Penner, J.E. Uncertainty analysis for estimates of the first indirect aerosol effect. *Atmos. Chem. Phys.* **2005**, *5*, 2935–2948. [CrossRef]
8. Stocker, T. *Climate Change 2013: The Physical Science Basis: Working Group I Contribution to the Fifth Assessment Report of the Intergovernmental Panel on Climate Change*; Cambridge University Press: New York, NY, USA, 2014; pp. 33–50.
9. Lolli, S.; Alparone, L.; Garzelli, A.; Vivone, G. Haze Correction for Contrast-Based Multispectral Pansharpening. *IEEE Geosci. Remote. Sens. Lett.* **2017**, *14*, 2255–2259. [CrossRef]
10. He, Q.Q.; Zhang, M.; Huang, B.; Tong, X.L. MODIS 3 km and 10 km aerosol optical depth for China: Evaluation and comparison. *Atmos. Environ.* **2017**, *153*, 150–162. [CrossRef]
11. Talbi, A.; Kerchich, Y.; Kerbachi, R.; Boughedaoui, M. Assessment of annual air pollution levels with PM1, PM2.5, PM10 and associated heavy metals in Algiers, Algeria. *Environ. Pollut.* **2018**, *232*, 252–263. [CrossRef]
12. Kim, K.H.; Kabir, E.; Kabir, S. A review on the human health impact of airborne particulate matter. *Environ. Int.* **2015**, *74*, 136–143. [CrossRef] [PubMed]
13. Oanh, N.T.K.; Upadhyaya, N.; Zhuang, Y.H.; Hao, Z.P.; Murthy, D.V.S.; Lestari, P.; Villarin, J.T.; Chengchua, K.; Co, H.X.; Dung, N.T.; et al. Particulate air pollution in six Asian cities: Spatial and temporal distributions, and associated sources. *Atmos. Environ.* **2006**, *40*, 3367–3380. [CrossRef]
14. Huebert, B.J.; Bates, T.; Russell, P.B.; Shi, G.Y.; Kim, Y.J.; Kawamura, K.; Carmichael, G.; Nakajima, T. An overview of ACE-Asia: Strategies for quantifying the relationships between Asian aerosols and their climatic impacts. *J. Geophys. Res. Atmos.* **2003**, *108*. [CrossRef]
15. Basart, S.; Perez, C.; Cuevas, E.; Baldasano, J.M.; Gobbi, G.P. Aerosol characterization in Northern Africa, Northeastern Atlantic, Mediterranean Basin and Middle East from direct-sun AERONET observations. *Atmos. Chem. Phys.* **2009**, *9*, 8265–8282. [CrossRef]
16. Agacayak, T.; Kindap, T.; Unal, A.; Pozzoli, L.; Mallet, M.; Solmon, F. A case study for Saharan dust transport over Turkey via RegCM4.1 model. *Atmos. Res.* **2015**, *153*, 392–403. [CrossRef]
17. Tegen, I.; Lacis, A.A.; Fung, I. The influence on climate forcing of mineral aerosols from disturbed soils. *Nature* **1996**, *380*, 419–422. [CrossRef]
18. Holben, B.N.; Eck, T.F.; Slutsker, I.; Tanre, D.; Buis, J.P.; Setzer, A.; Vermote, E.; Reagan, J.A.; Kaufman, Y.J.; Nakajima, T.; et al. AERONET—A federated instrument network and data archive for aerosol characterization. *Remote Sens. Environ.* **1998**, *66*, 1–16. [CrossRef]
19. Nichol, J.E.; Bilal, M. Validation of MODIS 3 km Resolution Aerosol Optical Depth Retrievals Over Asia. *Remote Sens.-Basel* **2016**, *8*, 328. [CrossRef]
20. Sayer, A.M.; Hsu, N.C.; Bettenhausen, C.; Jeong, M.J.; Holben, B.N.; Zhang, J. Global and regional evaluation of over-land spectral aerosol optical depth retrievals from SeaWiFS. *Atmos. Meas. Tech.* **2012**, *5*, 1761–1778. [CrossRef]
21. Kahn, R.A.; Gaitley, B.J.; Garay, M.J.; Diner, D.J.; Eck, T.F.; Smirnov, A.; Holben, B.N. Multiangle Imaging SpectroRadiometer global aerosol product assessment by comparison with the Aerosol Robotic Network. *J. Geophys. Res. Atmos.* **2010**, *115*. [CrossRef]
22. Vidot, J.; Santer, R.; Aznay, O. Evaluation of the MERIS aerosol product over land with AERONET. *Atmos. Chem. Phys.* **2008**, *8*, 7603–7617. [CrossRef]
23. Jackson, J.M.; Liu, H.Q.; Laszlo, I.; Kondragunta, S.; Remer, L.A.; Huang, J.F.; Huang, H.C. Suomi-NPP VIIRS aerosol algorithms and data products. *J. Geophys. Res. Atmos.* **2013**, *118*, 12673–12689. [CrossRef]
24. Remer, L.A.; Mattoo, S.; Levy, R.C.; Munchak, L.A. MODIS 3 km aerosol product: algorithm and global perspective. *Atmos. Meas. Tech.* **2013**, *6*, 1829–1844. [CrossRef]
25. Bilal, M.; Nichol, J.E. Evaluation of MODIS aerosol retrieval algorithms over the Beijing-Tianjin-Hebei region during low to very high pollution events. *J. Geophys. Res. Atmos.* **2015**, *120*, 7941–7957. [CrossRef]
26. Sayer, A.M.; Hsu, N.C.; Bettenhausen, C.; Jeong, M.J. Validation and uncertainty estimates for MODIS Collection 6 "Deep Blue" aerosol data. *J. Geophys. Res. Atmos.* **2013**, *118*, 7864–7872. [CrossRef]

27. Tao, M.H.; Chen, L.F.; Wang, Z.F.; Tao, J.H.; Che, H.Z.; Wang, X.H.; Wang, Y. Comparison and evaluation of the MODIS Collection 6 aerosol data in China. *J. Geophys. Res. Atmos.* **2015**, *120*, 6992–7005. [CrossRef]
28. Tian, X.P.; Liu, Q.; Li, X.H.; Wei, J. Validation and Comparison of MODIS C6.1 and C6 Aerosol Products over Beijing, China. *Remote Sens.-Basel* **2018**, *10*, 2021. [CrossRef]
29. Smirnov, A.; Holben, B.N.; Eck, T.F.; Dubovik, O.; Slutsker, I. Cloud-screening and quality control algorithms for the AERONET database. *Remote Sens. Environ.* **2000**, *73*, 337–349. [CrossRef]
30. Schaap, M.; Timmermans, R.M.A.; Koelemeijer, R.B.A.; de Leeuw, G.; Builtjes, P.J.H. Evaluation of MODIS aerosol optical thickness over Europe using sun photometer observations. *Atmos. Environ.* **2008**, *42*, 2187–2197. [CrossRef]
31. Li, D.; Qin, K.; Wu, L.X.; Xu, J.; Letu, H.S.; Zou, B.; He, Q.; Li, Y.F. Evaluation of JAXA Himawari-8-AHI Level-3 Aerosol Products over Eastern China. *Atmosphere-Basel* **2019**, *10*, 215. [CrossRef]
32. van Donkelaar, A.; Martin, R.V.; Brauer, M.; Boys, B.L. Use of Satellite Observations for Long-Term Exposure Assessment of Global Concentrations of Fine Particulate Matter. *Environ. Health Persp.* **2015**, *123*, 135–143. [CrossRef] [PubMed]
33. Hubanks, P.A. *MODIS Atmosphere QA Plan for Collection 005*; NASA Goddard Space Flight Center: Greenbelt, MD, USA, 2005; Volume 57.
34. Levy, R.C.; Mattoo, S.; Munchak, L.A.; Remer, L.A.; Sayer, A.M.; Patadia, F.; Hsu, N.C. The Collection 6 MODIS aerosol products over land and ocean. *Atmos. Meas. Tech.* **2013**, *6*, 2989–3034. [CrossRef]
35. Gupta, P.; Levy, R.C.; Mattoo, S.; Remer, L.A.; Munchak, L.A. A surface reflectance scheme for retrieving aerosol optical depth over urban surfaces in MODIS Dark Target retrieval algorithm. *Atmos. Meas. Tech.* **2016**, *9*, 3293–3308. [CrossRef]
36. Remer, L.A.; Kaufman, Y.J.; Tanre, D.; Mattoo, S.; Chu, D.A.; Martins, J.V.; Li, R.R.; Ichoku, C.; Levy, R.C.; Kleidman, R.G.; et al. The MODIS aerosol algorithm, products, and validation. *J. Atmos. Sci.* **2005**, *62*, 947–973. [CrossRef]
37. Hsu, N.C.; Jeong, M.J.; Bettenhausen, C.; Sayer, A.M.; Hansell, R.; Seftor, C.S.; Huang, J.; Tsay, S.C. Enhanced Deep Blue aerosol retrieval algorithm: The second generation. *J. Geophys. Res. Atmos.* **2013**, *118*, 9296–9315. [CrossRef]
38. Sayer, A.M.; Munchak, L.A.; Hsu, N.C.; Levy, R.C.; Bettenhausen, C.; Jeong, M.J. MODIS Collection 6 aerosol products: Comparison between Aqua's e-Deep Blue, Dark Target, and "merged" data sets, and usage recommendations. *J. Geophys. Res. Atmos.* **2014**, *119*, 13965–13989. [CrossRef]
39. Ichoku, C.; Chu, D.A.; Mattoo, S.; Kaufman, Y.J.; Remer, L.A.; Tanre, D.; Slutsker, I.; Holben, B.N. A spatio-temporal approach for global validation and analysis of MODIS aerosol products. *Geophys. Res. Lett.* **2002**, *29*. [CrossRef]
40. Xie, Y.; Zhang, Y.; Xiong, X.X.; Qu, J.J.; Che, H.Z. Validation of MODIS aerosol optical depth product over China using CARSNET measurements. *Atmos. Environ.* **2011**, *45*, 5970–5978. [CrossRef]
41. Bilal, M.; Nazeer, M.; Nichol, J.E.; Bleiweiss, M.P.; Qiu, Z.F.; Jakel, E.; Campbell, J.R.; Atique, L.; Huang, X.L.; Lolli, S. A Simplified and Robust Surface Reflectance Estimation Method (SREM) for Use over Diverse Land Surfaces Using Multi-Sensor Data. *Remote Sens.-Basel* **2019**, *11*, 1344. [CrossRef]
42. Levy, R.C.; Remer, L.A.; Kleidman, R.G.; Mattoo, S.; Ichoku, C.; Kahn, R.; Eck, T.F. Global evaluation of the Collection 5 MODIS dark-target aerosol products over land. *Atmos. Chem. Phys.* **2010**, *10*, 10399–10420. [CrossRef]
43. Mhawish, A.; Banerjee, T.; Broday, D.M.; Misra, A.; Tripathi, S.N. Evaluation of MODIS Collection 6 aerosol retrieval algorithms over Indo-Gangetic Plain: Implications of aerosols types and mass loading. *Remote Sens. Environ.* **2017**, *201*, 297–313. [CrossRef]
44. van der Werf, G.R.; Randerson, J.T.; Giglio, L.; Collatz, G.J.; Mu, M.; Kasibhatla, P.S.; Morton, D.C.; DeFries, R.S.; Jin, Y.; van Leeuwen, T.T. Global fire emissions and the contribution of deforestation, savanna, forest, agricultural, and peat fires (1997-2009). *Atmos. Chem. Phys.* **2010**, *10*, 11707–11735. [CrossRef]
45. Zhang, L.; Liao, H.; Li, J.P. Impacts of Asian summer monsoon on seasonal and interannual variations of aerosols over eastern China. *J. Geophys. Res. Atmos.* **2010**, *115*. [CrossRef]

MDPI
St. Alban-Anlage 66
4052 Basel
Switzerland
Tel. +41 61 683 77 34
Fax +41 61 302 89 18
www.mdpi.com

Atmosphere Editorial Office
E-mail: atmosphere@mdpi.com
www.mdpi.com/journal/atmosphere